Around the World in 18 Elements

Around the World in 18 Elements

David A. Scott
The King's School Canterbury, UK
Email: das@kings-school.co.uk

ISBN: 978-1-84973-804-0

A catalogue record for this book is available from the British Library

Published by The Royal Society of Chemistry,
Thomas Graham House, Science Park, Milton Road,
Cambridge CB4 0WF, UK

Registered Charity Number 207890

Visit our website at www.rsc.org/books

Printed in the United Kingdom by CPI Group (UK) Ltd, Croydon, CR0 4YY

For Sally

Author's Note

This book stems from an article on phosphorus published in the Association of Science Education's (ASE) journal *School Science Review*, volume 93, issue no. 344, March 2012. In the original article I attempted to look at different ways of re-engaging students with material on an A level chemistry syllabus as part of a revision program. From that comes this book of eighteen chapters, each on one element, with A level type questions peppered throughout. I don't pretend to offer anything greatly original here; all of the chemistry is in the public domain. I am merely trying to offer a slightly more informal approach, whilst slipping in questions that would be relevant to a student preparing for an A level in chemistry.

Current and future developments to post-sixteen chemistry courses in England and beyond are looking to provide more challenging questions for students and I have tried to keep this in mind when putting together the questions. Some questions are straightforward, others, I hope, take things into slightly new territory. It is important, though probably self-evident, to say what this book is NOT. It is not a textbook and, thus, students opening this book before they have covered most of the A level (or equivalent) syllabus may be put off (which is precisely the opposite of my intention). It also does not attempt to offer comprehensive syllabus coverage. I would hope that the book could be used by teachers looking for some new contexts and even students preparing for first year university courses for which a modicum of chemistry is required.

Around the World in 18 Elements
By David A. Scott

Published by the Royal Society of Chemistry, www.rsc.org

The eighteen elements I have chosen are not a particularly representative set of the elements in the periodic table. There are nine metals and nine non-metals but, as 80% of the elements in the periodic table are metals, this is clearly a disproportionately biased selection in favour of the non-metals. All I can say is that it reflects my own interests. I flirted with chapters on potassium, molybdenum, silver, gold, magnesium and silicon but plumped in the end for the eighteen detailed herein.

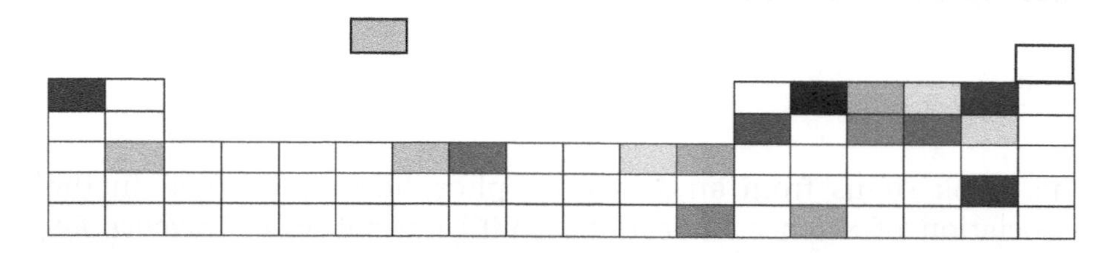

FURTHER READING

Rather than attempt to acknowledge the various writers that have fed into these chapters to a greater or lesser extent, I thought I'd take this opportunity to suggest some further reading, so here goes:

1. *The Periodic Table* by Primo Levi
2. *The Natural Selection of the Chemical Elements* by R. J. P Williams and J. J. R. Frausto da Silva
3. *Nature's Building Blocks* by John Emsley. (Basically anything by Emsley!)
4. *Gaia: A New Look at Life on Earth* by James Lovelock.
5. *Evolution's Destiny* by R. J. P Williams and R. E. M. Rickerby.

I would like to acknowledge gratitude to Geoff Auty, editor at the ASE's journal *School Science Review* for help and guidance with the phosphorus article in the March 2012 edition of the journal.

Preface

Of all the science A levels, it is chemistry that is often seen to be the "means to an end" science A level; for the medics and the vets it is non-negotiable, the bio-scientists know that, with the possible exception of ecology, they will be at a great disadvantage without it. Then there are the geologists, the material scientists, the environmental scientists and, of late, the forensic scientists too. Only occasionally, in my experience, does a pupil sidle up, look furtively left and right, and express *sotto voce* a desire to study chemistry at university. I think that one of the reasons for this is that an awful lot of chemistry is presented in the abstract rather than a specific context. To a certain extent, this is understandable as the nature of chemistry means that the contexts to which it is applicable are multiple and varied but this does not mean that contexts can't be given. As a biochemist, the context for me was always about what made chemistry "come alive" and, therefore, the applicability of organic chemistry is immediately obvious, but then so are the key concepts of physical chemistry (entropy and energy), redox chemistry (electron transfer reaction), transition metal chemistry (at the heart of so many enzymes) and so on. I can still remember learning about the Haber process for my own chemistry "O" level and wondering why we were focusing on such a seemingly unspectacular reaction. Had I been told that, according to some estimates, some 40% of the world's population would not be alive without it then I might have seen the relevance. I hope then that the eighteen chapters that follow are an accessible read, and that there are a few

Around the World in 18 Elements
By David A. Scott

Published by the Royal Society of Chemistry, www.rsc.org

questions that will make you think. I also hope that a few of the historical contexts will illustrate how the scientific enterprise moves on amidst the changing ideologies, political struggles and constant social turmoil that provides the context for that enterprise.

David A. Scott
Canterbury

Contents

Around the World in 18 Elements
By David A. Scott

Published by the Royal Society of Chemistry, www.rsc.org

CHAPTER 1

Phosphorus

Phosphorus has got itself a bit of a bad name over the years. According to Issac Asimov in his very readable *New Guide to Science*,[1] phosphorus was the impurity responsible for scuppering Henry Bessemer's initial attempts to produce high quality steel. It was also responsible for the awful medical condition known as "phossy jaw", which developed as a result of early "strike anywhere" matches that contained white phosphorus and, as we will see, this element also played a devastating part in World War II. It wasn't until the second half of the twentieth century that scientists began to get a full insight into just how vital phosphorus is for life.

1.1 PHOSPHORUS THE NON-METAL?

Phosphorus is the second member of group V in the periodic table, just below nitrogen. Its position firmly to the right of the stepped dividing line that all year 10 students draw on their periodic table places it firmly in the non-metals section of the table. However, at A level it is important to recognize that the material world rarely conforms to our hard and fast metaphysical categories, and that this line of demarcation is rather more "fuzzy" than implied at GCSE. The so-called metalloids highlight this more gradual change from metallic to non-metallic character; the semi-conducting properties of silicon and germanium are well known, but to find that phosphorus can exist

Around the World in 18 Elements
By David A. Scott

Published by the Royal Society of Chemistry, www.rsc.org

in an allotropic form that can conduct electricity (black phosphorus) came as quite a shock...at least to this author!

Lesson 1: be prepared to appreciate the subtlety of chemistry. The other more common allotropes of phosphorus are white and red.

Q1. Define the term allotrope and name one other non-metal element and one metal element that can exist as allotropes.

Q2. A typical human body contains between 600 to 800 g of phosphorus and yet a dosage of 100 mg of white phosphorus can be fatal. How can you explain this *apparent* contradiction?

1.2 THE DISCOVERY OF PHOSPHORUS

The first isolation of phosphorus is attributed to Henning Brandt (Hamburg, 1669). The story is well known for its "yuck" factor in that Brandt required gallons of human urine, acquired from Hamburg bierkellers, which he then commenced to heat, evaporate, allow to putrefy for days, filter and heat again until samples of phosphorus vapor were condensed over and collected under water. However, at this time in history, the alchemical paradigm held sway and Brandt thought that he might have stumbled on the fabled philosopher's stone: that which could turn base metal into gold. Brandt failed to make his fortune but was, at least, spared the horrifying epilogue to the story of Hamburg and phosphorus. The use of white phosphorus in the firebombing of Hamburg in 1943 was estimated to have killed between 50 000–100 000 men, women and children.

Q3.

a) Calculate the number of moles of phosphorus atoms in a 14 kg bomb.

b) Write a balanced equation for the complete combustion of phosphorus. You can assume white phosphorus exists as P_4 molecules and that the only combustion product is P_4O_{10}.

c) Given that the $\Delta H^{\theta}{}_{f}$ for P_4O_{10} is −2984 kJ mol^{-1}, calculate the total amount of heat energy released by a 14 kg phosphorus bomb.

Another allotrope of phosphorus, as yet unmentioned, is the diatomic P_2 molecule. Sitting below nitrogen it is perhaps not surprising that P_2 molecules are linked by triple bonds ($P{\equiv}P$).

Q4.

a) Draw a dot and cross diagram of the $P_{2(g)}$ molecule.
b) The mean bond enthalpy for a single P–P bond is 198 kJ mol^{-1} and the bond enthalpy for P≡P is 489 kJ mol^{-1}. Using a Hess's law cycle, calculate the enthalpy change for the following reaction:

$$P_{4(g)} \rightarrow 2P_{2(g)} \quad (1.1)$$

c) Using your answer to b) suggest which is the more stable allotrope under standard conditions.

We will return to this equation later in Q11.

No less incendiary, but rather less destructive, is the role of phosphorus in matches. Tetraphosphorus trisulfide (P_4S_3) is the fuel component of "strike anywhere" matches. The oxidizing agent, potassium chlorate (V), is added to the match head and sufficient friction initiates the following reaction:

$$3P_4S_{3(s)} + 16KClO_{3(s)} \rightarrow 16KCl_{(s)} + 9SO_{2(g)} + 3P_4O_{10(s)} \quad (1.2)$$

Q5. Calculate the mass of potassium chlorate (V) that would be required to completely combust 1 kg of P_4S_3 according to eqn 1.2.

1.3 OXIDATION STATES OF PHOSPHORUS

It is clear from eqn 1.2 that phosphorus is oxidized in the reaction. However, the concept of the oxidation number is too useful to be kept only for reactions involving oxygen. With an electronegativity value of 2.1, phosphorus might be expected to form compounds with a range of formal oxidation states or numbers.

Q6. Assign an oxidation number to phosphorus in each of the following compounds (AS level):

a) P_4O_6
b) Ca_3P_2
c) H_3PO_4
d) Na_2HPO_3

A comparison of two phosphide compounds (containing the P^{3-} ion) illustrates a trend in **bond character.** Sodium phosphide (Na_3P) is an ionic black salt that is insoluble in water. Aluminium phosphide (AlP) hydrolyses on contact with water to produce highly toxic phosphine gas (PH_3).

Q7. Explain why aluminium phosphide has a greater degree of covalent character than sodium phosphide.

$$AlP_{(s)} + 3H_2O_{(l)} \rightarrow Al(OH)_{3(s)} + PH_{3(g)} \qquad (1.3)$$

Q8.

a) Draw a dot and cross diagram of the PH_3 molecule, clearly showing the shape of the molecule.
b) Phosphine ($M_r = 34$ g mol^{-1}) has a boiling point of -88 °C. Ammonia ($M_r = 17$ g mol^{-1}) has a boiling point of -33 °C. Account for these differences in the boiling points.

Like nitrogen, phosphorus can show oxidation states of -3 and + 3, but the ability of phosphorus to expand its octet allows a greater chemical diversity.

Q9.

a) Write out the s,p,d electronic notation for a phosphorus atom.
b) Phosphorus trichloride will react further with chlorine to produce phosphorus pentachloride. Write out a balanced equation for this reaction.
c) Show how sp^3d hybridization allows $PCl_{5(g)}$ to form.
d) Draw a dot and cross diagram of PCl_5 and label two **different** bond angles.

1.4 THE PHOSPHORUS CYCLE

Like its periodic table neighbors, carbon, nitrogen and sulfur, phosphorus forms **acidic oxides** but, unlike them, it does not form gaseous oxides and thus does not enter the atmosphere in any significant quantities. This has considerable implications for the cycling of phosphorus on the Earth's surface, where it resides in rocks or dissolved in rivers and oceans. Issac Asimov had this to say about phosphorus:

"We may be able to substitute nuclear power for coal, and plastics for wood, and yeast for meat, and friendliness for isolation—but for phosphorus there is neither substitute nor replacement."[2]

Just how important phosphorus is to life can be appreciated by considering the graph taken from Chopra and Lineweaver (Figure 1.1).[3]

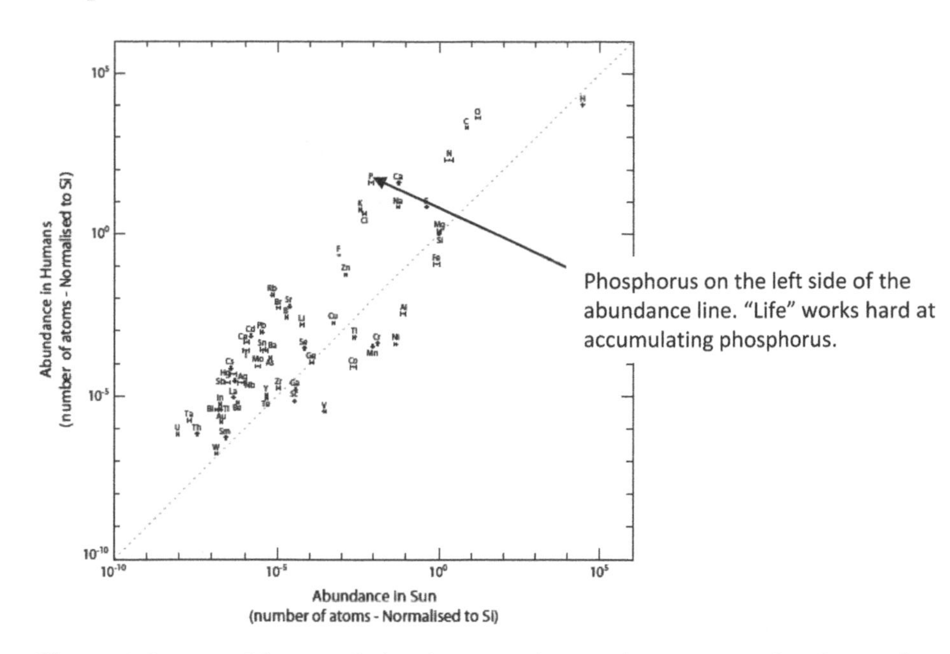

Figure 1.1 A positive correlation between elemental percentage abundances (by number of atoms), as represented by humans (y axis) and the devolatised Sun (x axis). The abundances are normalized to silicon. Reproduced with permission of the authors, Chopra *et al.*[3]

The significance of phosphorus lying to the upper left hand side of the dashed diagonal shows that all life concentrates phosphorus within itself to a higher degree than any other element. To put it another way: for any one atom of phosphorus in the universe at large, there are about 10 000 phosphorus atoms in known living systems.

Because phosphorus forms no gaseous oxides, its cycling is determined by the Earth's geological and hydrological cycles, which might be thought to limit its availability, and yet its importance to life can hardly be overstated! Consider the following examples:

i) The HPO_4^{2-} ion is vital in buffering blood serum pH.
ii) In phospholipids it is a major constituent of cell membranes.
iii) It forms part of the backbone of RNA and DNA molecules.
iv) In adenosine triphosphate (ATP), the "universal" energy molecule, it is present no less than three times.
v) It can be found in cyclic adenosine monophosphate (AMP), nicotinamide adenine dinucleotide phosphate (NADPH), guanosine triphosphate (GTP)...the list goes on!

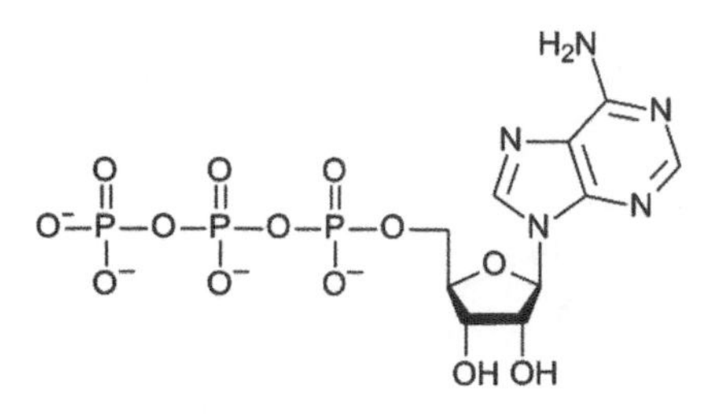

Figure 1.2 The structure of ATP.

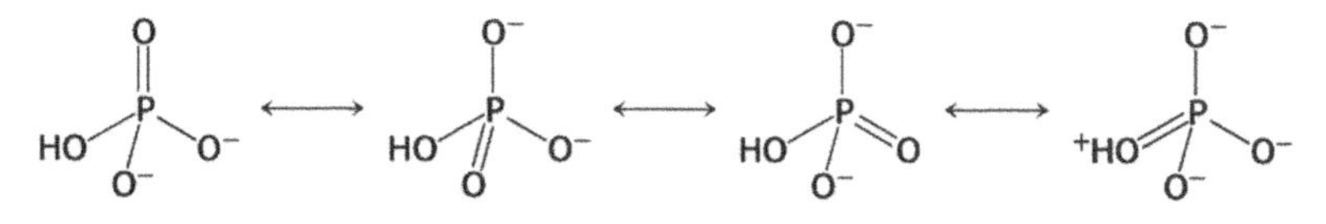

Figure 1.3 Resonance structures for HPO_4^{2-}.

It is worth dealing with a common misconception at this juncture: in chemistry lessons students are often told that bond formation is exothermic and bond breaking is endothermic. All too frequently they are then told in biology that the universal energy "unit" is generated by ATP breaking to form adenosine diphosphate (ADP) + phosphate. The hydrolysis (*i.e.*, breaking) of the P–O bond between the second and third phosphate group is **endothermic.** However, the resultant hydrogen phosphate ion (HPO_4^{2-}) then forms a resonance-stabilized structure, which is an exothermic process. You may be familiar with the consideration of the resonance stability in benzene, which makes it less reactive and thus more energetically stable than the hypothetical cyclo-1,3,5,-hexatriene (Figures 1.2 and 1.3).

The **net** result of this process has a $\Delta H_r = -30.5$ kJ mol^{-1} under physiological conditions. This energy can then be used to drive reactions that are thermodynamically unfavorable, such as protein synthesis.

Once released into the cytoplasm the HPO_4^{2-} has an important role to play in buffering pH. Phosphoric (V) acid has the formula H_3PO_4. However, in the neutral conditions of the cytoplasm the following equilibrium in eqn 1.4 is important:

$$H_2PO_4{}^-{}_{(aq)} \rightleftharpoons HPO_4{}^{2-}{}_{(aq)} + H^+{}_{(aq)} \tag{1.4}$$

The reversibility of this reaction means that small additions of H^+ or ^-OH can be buffered so as to retain the cytoplasmic pH at 7.0.

Q10.

a) Write an expression for K_a for the $H_2PO_4^-$ ion acting as an acid.
b) Given that the K_a value for eqn 1.4 is 6.2×10^{-8} mol dm^{-3}, calculate the ratio of $H_2PO_4^-$: HPO_4^{2-} at pH = 7.0

1.5 WILL PHOSPHORUS EVER RUN OUT?

Given that one of the first principles of chemistry is that "matter is never created or destroyed in a chemical reaction", akin to the first law of thermodynamics, which tells us that energy can never be consumed, the idea that phosphorus can be "used up" is nonsense. It should be more correctly stated that sources of phosphorus that are *economically viable to extract* can be over-consumed. Many of my students get very excited when I tell them that there are 10 million tonnes of pure gold in the world's oceans but, at a concentration of 10 parts per 1×10^{12}, extracting the gold would cost massively more than the value of the gold extracted. However, the value of a raw material is also not fixed; for example, the price per tonne of phosphorus in 1993 was $21.38, however, by 2008 this had climbed to $113.00 per tonne.[6] Clearly, a greater than five-fold increase in the price per tonne can affect the viability of a mineral resource.

We will now return to Q4 and reconsider the feasibility of the reaction shown in eqn 1.1:

$$P_{4(g)} \rightarrow 2P_{2(g)} \tag{1.1}$$

Hopefully, you will have concluded that under standard conditions (1 atm. and 298 K) the reaction in eqn 1.1 is **not** feasible and yet, under certain conditions, it becomes feasible. To show this we must recall the entropy equations shown in eqn 1.5 and 1.6.

$$\Delta S_{TOTAL} = \Delta S_{SYSTEM} + \Delta S_{SURROUNDINGS} \tag{1.6}$$

and that

$$\Delta S_{SURROUNDINGS} = -\Delta H/T \tag{1.5}$$

Q11. Use the data given in Table 1.1 to calculate the temperature at which the conversion of P_4 to $2P_2$ becomes feasible.

Table 1.1 Standard enthalpies of formation and standard entropy values for P_2 and P_4.

Substance	S^{θ}/J mol^{-1} K^{-1}	ΔH^{θ}_f/kJ mol^{-1}
P_2	218	144
P_4	41	0

1.6 PHOSPHORUS EXTRACTION AND "GREEN CHEMISTRY"

One of the more pressing concerns in the industrial chemistry sector these days is how to produce the desired product in the most energy-efficient way and to do so with the optimum *atom economy*. The concept of atom economy is a relatively recent one. It is defined as follows in eqn 1.7:

$$\text{Atom economy} = \frac{\text{molecular mass of desired product}}{\text{molecular mass of all reactants}} \times 100 \qquad (1.7)$$

The importance of atom economy has grown as regulations involving the processing of waste materials have become more stringent and thus it has become less economically and environmentally viable to simply dump waste.

The extraction of phosphorus from a naturally occurring compound, such as calcium phosphate, is represented in eqn 1.8:

$$Ca_3(PO_4)_{2(s)} + 10C_{(s)} + 6SiO_{2(s)} \rightarrow 6CaSiO_{3(l)} + 10CO_{(g)} + P_{4(g)} \qquad (1.8)$$

This reaction is carried out in an electric furnace at 1400 °C.

Q12.

a) Calculate the atom economy for this reaction.
b) How might the economic viability of this reaction be improved?

With geological resources for phosphorus becoming less economically viable, chemists are now developing methods of extracting phosphorus from animal and even human waste.

1.7 CONCLUSIONS

In the *The Periodic Table* by Primo Levi,[4] chapter 9 is titled "Phosphorus". "...It is not an emotionally neutral element," says Levi. As the "bringer of light", we have seen it can be the harbinger of destruction or the very spark of life. It may be that phosphorus availability is the key rate-limiting element that determines the development of life itself and yet, just as we are getting comfortable with the idea that phosphorus may indeed be life's "bottleneck", up pops GFAJ-1, a bacterial resident of Mono lake in California that appears to be able to substitute arsenic (As) for phosphorus in its DNA. Although it should be noted that there is still a good deal of controversy regarding the reliability of the evidence of arsenic being incorporated into GFAJ-1's DNA. That said, we should perhaps heed lesson 1: "be prepared to appreciate the subtlety of chemistry".

REFERENCES

1. I. Asimov, *Asimov's New Guide to Science*, 1984, Penguin, London, pp. 278–279.
2. I. Asimov, *A Short History of Chemistry*, 1979, Anchor Books, Garden City, New York.
3. A. Chopra, C. H. Lineweaver, J. J. Brocks. T. R. Ireland, Paleo-ecophylostoichiemetrics: Searching for the elemental composition of the last universal common ancestor, *Proceedings of the 9th Space Science Conference*, Sydney, 20–30th Sept 2009, National Space Society of Australia, Sydney.
4. P. Levi, *The Periodic Table*, 1986, Abacus, London, p. 120.

Answers to the questions in this chapter can be found on pages 168–171.

CHAPTER 2

Iron

In Douglas Adams' *Hitchhiker's Guide to the Galaxy*,[1] Arthur Dent (Earthman) is extremely put out when he discovers that Ford Prefect (his friend, but not in fact a human) has changed the Guide's entry on Earth from "harmless" to "mostly harmless". Had Ford given a similarly perfunctory chemical analysis of Earth he might well have entered "mostly iron silicate".

The complexity of life on Earth is certainly dependent upon carbon, liquid water and a source of energy, but the role that iron plays in life on Earth is as varied as it is vital as I hope to convince you.

It is impossible to overstate the significance of iron to both our culture and life on Earth in general. So fundamental is its association that iron has been connected with "masculine" attributes in alchemy and is still used figuratively today to give "cast iron" guarantees. It has been known since antiquity and thus, much to the irritation of today's chemistry students, gets its symbol (Fe) from the Latin *ferrum* rather than the more systematic Ir, which is reserved for its transition metal neighbor, iridium. It is the 26th element in the periodic table and yet is the 6th most abundant element in the known universe and the 4th most abundant in the Earth's crust. It is, in some cases quite literally, at the centre of things!

2.1 JOURNEY TO THE CENTRE OF THE EARTH

In 1878 Jules Verne wrote *Journey to the Centre of the Earth.*[2] Ever since Charles Lyell's 1863 *Geological Evidences of the Antiquity of Man*,[3] people had been less willing to accommodate the literal truth of

Around the World in 18 Elements
By David A. Scott

Published by the Royal Society of Chemistry, www.rsc.org

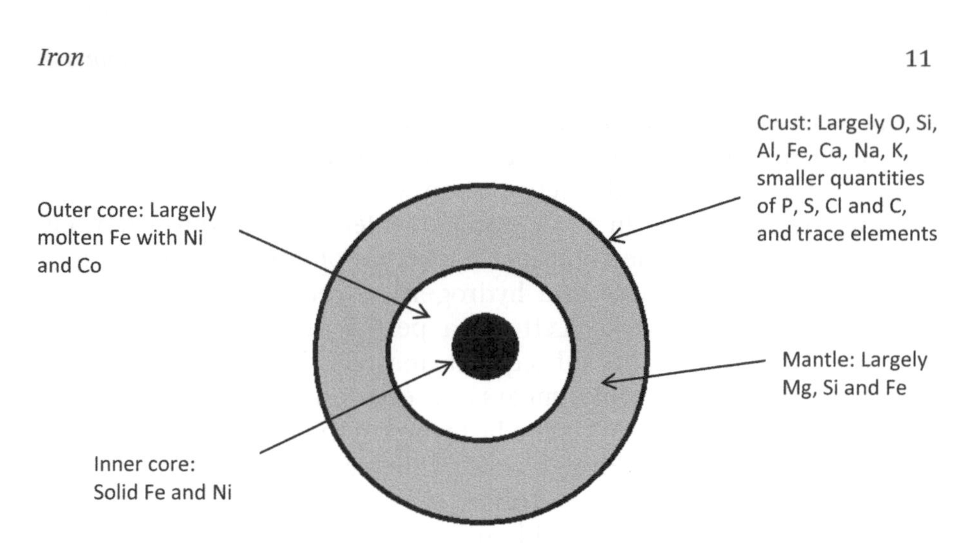

Figure 2.1 Illustration of the Earth's inner structure showing the principal elements present.

the Bible and scientific speculation about Man's place in the universe had blossomed. The received wisdom about the Earth's centre today is rather more prosaic than Verne's lost world of dinosaurs. The centre of the Earth is iron (Figure 2.1).

About 4.6 billion years ago the Earth formed from cosmic debris. Being relatively close to the Sun, gravitation ensured that the Earth would be formed from the heavier elements and, whilst still in its molten state, the heavier iron sank to what would become the Earth's centre. 4.6 billion years later the Earth is still geologically active and thus the Earth's mantle and crust are far from chemically homogenous. Volcanic activity has ensured that many minerals have been concentrated at various parts of the Earth's crust. Later, a plentiful supply of high quality iron ore, together with a good supply of coal, would prove to be a great benefit to the indigenous tribes to come. But we're getting ahead of ourselves—*life hasn't evolved yet!*

Q1. The Earth's mass is approximately 6.0×10^{27} g. If 70% of the Earth's mass is the inner and outer core, calculate the number of moles of iron in the Earth's core. For the purposes of the calculation, assume that the core is 55% iron.

2.2 LIFE ON EARTH

The origin of life on Earth is a mystery. The question is not made much easier by the fact that what constitutes "life" is still an open question. That said, few scientists today would hold with theories of "*vitalism*" (the belief in some additional and mysterious life force).

From a Darwinian point of view, once you have a chemical system that can replicate itself—albeit with the very occasional error—and a supply of materials and energy, then you only need time before complex life emerges. In 1953 Stanley Miller[4] carried out a famous experiment in which he subjected a mixture of gases (methane, ammonia, carbon monoxide and hydrogen), together with water, to a series of electrical sparks. Left for a period of days, the chemical mixture turned pink and subsequent analysis showed that the mixture contained amino acids, sugars and even nucleotides. It would be a mistake to think that this is the last word in the evolution of life but, if some of the basic building blocks of life can be generated in days, then what can be achieved in eons?

There is, however, one problem for these complex organic molecules formed at the Earth's surface: cosmic radiation. Whether from our Sun or from outside the solar system, cosmic rays are high energy particles (mainly protons) that can smash complex organic molecules to pieces. Whatever life is, it is certainly complex, and complex molecules don't last long in the shooting gallery of cosmic radiation. It is the Earth's iron core that is responsible for its magnetic field. The dynamic nature of the inner and outer core generates a magnetic field called the magnetosphere. This deflects cosmic rays, drawing them in at the Earth's poles. As these particles hurtle into the Earth's atmosphere at the poles, they interact with the gases in the atmosphere giving rise to the *Aurora borealis* at the North Pole and the *Aurora australis* at the South. The absence of the magnetosphere may well have meant that life on Earth would never have got started at all.

2.3 THE IRON AGE

Iron is the only element to have a historical age named after it. Copper had to suffer the indignity of being alloyed with tin to lodge itself into our consciousness as the Bronze Age. And yet the Iron Age was a long time coming! Copper is a more co-operative material. It is easily teased out of its ore and, when alloyed with tin, gives a weighty and versatile material that was used for weapons and ornaments. Iron is all together more recalcitrant stuff. Not only did it require higher temperatures to be extracted from its ore, but excess carburization on the one hand or incomplete reduction on the other repeatedly frustrated the early blacksmith. It may be that many early cultures preferred the dependable but heavy bronze sword to the lighter, harder but unpredictably brittle iron. Nevertheless, as the technology developed, it was iron and then steel that was to inherit the Earth.

2.4 OUR IRON-CLAD CIVILIZATION

In 1919, not long after the end of the mechanized horror that was World War I, Professor R.H. Whitbeck, writing in *Scientific Monthly*,[5] wrote:

> *"Mankind stepped from an era in which his highest material achievements were in stone structures – to the era of machines which multiply human energy, speed and ability in hundreds of ways"*

There can be little doubt that iron, alloyed in various ways to make materials ranging from the finest hair spring of a watch to armor-piercing projectiles, has profoundly shaped civilization.

"The world," says Whitbeck in conclusion, "has come under the domination of the peoples that have great reserves of coal and iron and know how to use them".

2.5 THE VERSATILITY OF IRON CHEMISTRY

In introducing the chemistry of iron to students, the unwary chemistry teacher may find him or herself launching into something terminally confusing like:

> *"Iron forms two ions, the iron II ion and the iron III ion too…"*

If, however, you manage to avoid this excruciating pitfall, then the chemistry of iron is indeed truly versatile. The two most common oxidation states of iron are iron II (Fe^{2+}) and iron III (Fe^{3+}). It is the ability of iron to shift between these two oxidation states that makes for some important chemistry.

Q2. Give the s,p,d notation of the following iron species: Fe, Fe^{2+} and Fe^{3+}.

Q3. Give oxidation numbers for iron and work out the % mass of iron in the following compounds: (i) Fe_2O_3, (ii) Fe_3O_4 and (iii) $FeCO_3$.

2.6 IRON AND THE GREAT OXIDATION EVENT

From the formation of the Earth about 4.6 billion years ago to about 2.4 billion years ago, Fe^{2+} ions were abundant in the oceans. The evolution of photosynthetic organisms, however, was to have a marked impact on iron availability in the oceans. Starting slowly, photosynthetic organisms started to produce diatomic oxygen as a by-product of glucose synthesis. As atmospheric oxygen built up, Fe^{2+} ions were oxidized to Fe^{3+} ions. Fe^{3+} ions are about 10^{12} times less soluble than Fe^{2+} ions and, consequently, iron precipitates out as iron III hydroxide. Banded iron formations (BIFs) are geological records of this event and provide the rich iron ore reserves that are exploited today.

Interestingly, analysis of lunar soil suggests an abundance of 0.5% *metallic* iron—not surprising given the absence of free oxygen on the moon. A future manned mission to Mars may well involve the manufacture of the vessel from lunar iron.

Q4. Balance the following equation and explain why it can be referred to as a *disproportionation reaction*:

$$\_\_FeO \rightarrow \_\_Fe_3O_4 + \_\_Fe$$

2.7 SOME IRON COMPOUNDS: MOHR'S SALT

With its ability to adopt different oxidation states, iron compounds have the potential to be useful in redox analysis. However, iron II sulfate is unstable and can be oxidized if exposed to air and, consequently, is unsuitable as a standard in volumetric analysis. K.F. Mohr (1806–1879) was a professor of pharmacy at the University of Bonn in Germany and one of his many contributions to chemistry was the double salt named after him. The advantage of the double salt was that it stabilized the Fe^{2+} ions, making them more resistant to oxidation. Mohr's salt has the formula: $(NH_4)_2SO_4 \cdot FeSO_4 \cdot 6H_2O$.

Q5.

a) Calculate the percentage mass of iron in Mohr's salt.
b) A standard solution 0.098 M Mohr's salt was used to calculate the concentration of a solution of $KMnO_4$. If 23.45 cm^3 of the standard solution was required to reach an end point with 25.00 cm^3 of the $KMnO_4$ solution, calculate the concentration of the $KMnO_4$ solution.
c) The M_r of Mohr's salt $= 392.0\,g\,mol^{-1}$ and the M_r of $FeSO_4 \cdot 7H_2O = 277.9$. Other than the stability of the Fe^{2+} ion, can you give another reason why Mohr's salt is a better standard solution for volumetric analysis.
d) Fe^{2+} ions are more stable in solutions of a lower pH. Using equations, explain why a solution containing $(NH_4)_2SO_4$ has a pH lower than 7.

2.8 IRON AT THE CENTRE OF THINGS AGAIN

The keen student preparing for GCSE chemistry will know the test for Fe^{2+} ions in solution:

1) Add sodium hydroxide solution.
2) The presence of Fe^{2+} ions is indicated by a grey/green precipitate of $Fe(OH)_2$.
3) After being left to stand for a few minutes, the top of the precipitate will turn a rust brown colour. This is due to the oxidation of Fe^{2+} ions to Fe^{3+}.

This reaction, subtly modified and made reversible is, in essence, the reaction that provides oxygen to our tissues for respiration. At the centre of a porphyrin ring (Fig. 2.2) is an Fe^{2+} ion.

Q6. Work out the molecular mass of the porphyrin ring shown in Fig. 2.2, with the Fe^{2+} ion at its centre.

Haemoglobin is a complex protein molecule present in red blood cells in which four of these rings are held in place. In areas of high oxygen partial pressure, oxygen molecules bond to the Fe^{2+} ions but in such a way that they can be released in environments of low oxygen partial pressure, such as muscles. Haemoglobin thus provides a sort of shuttle service, transferring oxygen from the lungs to the tissues where it is needed for respiration.

We have about 3.7 g of iron in our body, of which nearly 70% is found in haemoglobin. So important is iron's role in our body that a range of specialized proteins have evolved to store the extra iron

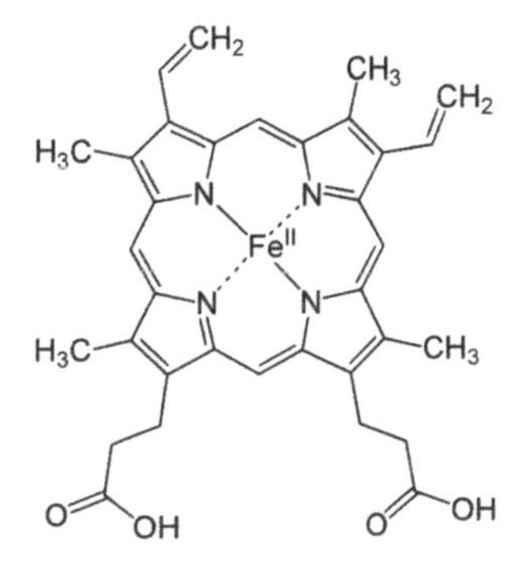

Figure 2.2 Fe^{2+} at the centre of a porphyrin ring.

(ferritins) and transport it to the required tissue when needed (transferrins). This is important because the reactivity of free Fe^{2+} ions in solution can result in the formation of radicals, as shown in eqn 2.1.

$$Fe^{2+} + H_2O_2 \rightarrow Fe^{3+} + {}^{-}OH + HO^{\bullet} \qquad (2.1)$$

Radicals are highly reactive species that can do considerable damage at the molecular level.

The importance of iron for life is highlighted by the existence of a class of proteins called transferrins—whose job it is to scavenge iron from the environment—and ferritins, whose job it is to store iron in the form of Fe^{3+}. (A typical ferritin protein of molecular mass 460 000 Da (1 Dalton = 1 atomic mass unit) may be up to 20% iron by mass. Iron-based enzymes are also involved in nitrogen fixation.)

Q7. If a ferritin protein has a molecular mass of 460 000 Da and is 20% by mass iron, calculate the number of Fe^{3+} ions present per ferritin to three significant figures.

Q8. Using the following standard electrode potentials (SEPs, eqn 2.2 and 2.3), show that free oxygen will convert iron II to iron III under standard conditions.

$$Fe^{3+} + e^{-} \rightleftharpoons Fe^{2+} \quad E^{\theta} = +0.77 \qquad (2.2)$$

$$\tfrac{1}{2}O_2 + 2H^{+} + 2e^{-} \rightleftharpoons H_2O \quad E^{\theta} = +1.23 \qquad (2.3)$$

Q9. The percentage of iron in a 3.50 g piece of steel is analyzed by the following method:

1) The iron is reacted with 100 cm^3 of 2.00 M (excess) hydrochloric acid, converting all the metallic iron to Fe^{2+} ions.
2) The resultant solution is made up to 500 cm^3.
3) Titrations are performed against 25 cm^3 of this solution with 0.02 M $KMnO_{4(aq)}$.
4) The average end-point titre recorded is 29.75 cm^3.

a) Calculate the percentage of iron in the steel sample.
b) Give two assumptions that you have made in presenting your result in part a).

2.9 GLOBAL WARMING! IRON COMES TO THE RESCUE?

The causes of climate change will always be a bone of contention. For some, the complexity of modeling the problem will be reason enough to dispute any conclusions that are not to their liking. But there seems to be little doubt that it is changing and pretty rapidly at that. One of the ways that has been suggested to mitigate the build up of atmospheric carbon dioxide is ocean iron fertilization. A number of experiments have been carried out in which iron II sulfate has been added to the ocean surface waters. The subsequent phytoplankton bloom results in measurable carbon fixation and, by ultimately removing this fixed carbon to lower ocean depths, results in a form of carbon capture that may prove effective. However, it is fair to say that ocean iron fertilization is unlikely, on its own, to address the climate change problem. In addition, geochemical engineering on a large scale may have long-term implications as yet unforeseen. For a start, iron II ions, once oxidized, can have an impact on the pH of a solution.

Q10. In an aqueous solution, iron III ions exist in the following equilibrium:

$$[Fe(H_2O)_6]^{3+} + H_2O \rightleftharpoons [Fe(H_2O)_5OH]^{2+} + H_3O^+ \quad (2.4)$$

The pK_a value for this reaction is 2.2.

a) Calculate the pH of a 5×10^{-2} M solution of $Fe(NO_3)_3$. What assumptions have you made?
b) Explain why iron III carbonate does not exist.

2.10 CONCLUSIONS: IRON AND MARS

For the Romans, the God Mars was the god of war and, appropriately enough as we have seen, war, like Mars, was associated with the metal iron. What the Romans could never have known is that the visibly orange hue that Mars reflects is due to none other than iron III oxide. NASA has determined that Martian soil is between 5–14% iron oxide and thus, one day, martian blast furnaces may be a step towards opening up the solar system for further exploration.

The symbol for Mars, ♂, may turn out to be a prescient symbol representing mankind leaving the planet to explore the outer reaches of the solar system! (Although not necessarily in a Volvo.)

REFERENCES

1. D. Adams, *The Hitchhiker's guide to the Galaxy*, 1979, Pan Books, London.
2. J. Verne, *Journey to the Centre of the Earth*, Oxford University Press, Oxford, 1992.
3. C. Lyell, *Geological Evidences for the Antiquity of Man*, 1863, London: Published By J.M. Dent & Sons Ltd., and In New York By E.P. Dutton & Co.
4. S. Miller, *Science*, 1953, **117**(3046), 528.
5. R. H. Whitbeck, Our Iron Clad Civilization, *The Scientific Monthly*, 1919, **9**(2), 125.

Answers to questions in this chapter can be found on pages 172–173.

CHAPTER 3

Nitrogen

Not for the first time in this book, in nitrogen we meet a seemingly paradoxical element. To early chemists it was known as Azote, from the Greek for, literally, "without life". Once the active part of air is removed (the bit we now call oxygen), we are left with a lifeless gas: Azote. And yet not only is nitrogen essential for life, the nitro prefix—indicating the presence of nitrogen in a compound—seems to crop up in some of the most reactive of substances: nitroglycerin for a start! It is the *N* in TNT and the *nitro* in *nitro*cellulose (gun cotton), confusing!

Fritz Haber developed a method for activating nitrogen, enabling the mass production of explosives to help the German army in World War I, and yet his methods are now responsible for feeding nearly half of the world's population. Nitrogen has some questions to answer.

3.1 NITROGEN: LIFE-LIMITING ELEMENT

About 99% of living matter is composed of the elements carbon, hydrogen, oxygen and nitrogen. Of these four elements C, H and O are never in short supply here on Earth. Pure water vapour and carbon dioxide are present in the air and precipitation ensures that many plants have the necessary supply of fresh water. Nitrogen, however, is less easily coaxed out of the air. Despite making up over 78% of the air's composition, only bacteria and some archaea have managed to prize apart the N≡N molecule, making it available for other organisms.

Around the World in 18 Elements
By David A. Scott

Published by the Royal Society of Chemistry, www.rsc.org

Despite this rather narrow bottleneck, nitrogen is vital for proteins and nucleic acids. Prior to the large-scale production of synthetic nitrogen-based fertilizers, farmers had known for centuries that land needed to be left fallow if it was to recover from a previous year's crop growth of wheat or barley. Later, it became clear that certain plants would help the land recover its fertility and crop rotation was employed. Today, millions of tonnes of ammonium nitrate-based fertilizers ensure we get the maximum usage of arable land.

3.2 ACTIVATING NITROGEN

Nitrogen can react with oxygen according to eqn 3.1:

$$N_2 + O_2 \rightarrow 2NO \tag{3.1}$$

The bond enthalpies in Table 3.1 can be used to calculate the enthalpy of this reaction under standard conditions.

Q1.

a) Calculate the enthalpy of reaction for the reaction between nitrogen and oxygen to make nitrogen monoxide (eqn 3.1).
b) Calculate the standard entropy change for the same reaction.
c) Calculate the temperature at which the reaction in eqn 3.1 becomes thermodynamically feasible.
d) Do your answers to a, b and c tell you anything about the rate of the reaction? (The clue is in the products!)

A very bright student of mine once lamented that just as you get to grips with the "rules" in chemistry, "along comes an important exception". There's probably something in this; and for the strong maths and physics students, the exceptions seem to prove particularly frustrating. I always try to convince them that it is the exceptions that make chemistry interesting. It can be the exceptions that *really test* our theories, or maybe even overturn them. Let's have a look at some exceptions in the next setions.

Table 3.1 Bond enthalpies and entropies of N_2, O_2 and NO.

Molecule	*Bond enthalpy/kJ mol^{-1}*	*Entropy/J mol^{-1} K^{-1}*
N≡N	945	96
O=O	496	103
N=O	631	211

3.3 RATES OF REACTION

The rate of a chemical reaction always speeds up as the temperature increases. . .except when it doesn't!

The rate of reaction between nitrogen monoxide and oxygen is an exception, and it does test our understanding of reaction kinetics. The balanced stoichiometric equation is shown in eqn 3.2:

$$2NO_{(g)} + O_{2(g)} \rightarrow 2NO_{2(g)} \quad (3.2)$$

Q2.

a) Explain why this reaction is extremely unlikely to take place in one step.

We can deduce nothing about the reaction mechanism from the stoichiometric equation. For that we need to generate some experimental data (Table 3.2).

The rate equation for this reaction can be written as follows:

$$\text{Rate} = k[NO]^x[O_2]^y \quad (3.3)$$

b)

(i) Use the data in Table 3.2 to determine the values of x and y in eqn 3.3.
(ii) Calculate a value for the rate constant k in the rate equation (eqn 3.3) and give its units.

All textbook stuff so far! But here's the interesting bit: if the reactions are repeated at 473 K, the value of the rate constant **approximately halves!**

(iii) Suggest a reaction mechanism that is consistent with all of the information given above. Can you give a plausible explanation for the rate of this reaction **decreasing** at a higher temperature?

Table 3.2 Experimental data for the reaction shown in eqn 3.2.

Experiment number[a]	*Initial [NO]/mol dm^{-3}*	*Initial [O$_2$]/mol dm^{-3}*	*Initial rate/mol dm^{-3} s^{-1}*
1	0.0222	0.00866	0.0222
2	0.0440	0.00866	0.0902
3	0.0440	0.01800	0.180

[a]All reactions carried out at 298K.

3.4 THE LAW OF CONSTANT COMPOSITION

The law of constant composition states that all samples of a given chemical compound have the same elemental composition by mass.

Were you to try and verify this theory using the composition of oxygen and nitrogen in "nitrogen oxide", you would have made a very bad choice! Here are some oxides of nitrogen:

$$N_2O,\ NO,\ NO_2,\ N_2O_3,\ N_2O_4,\ N_2O_5$$

Not only are there five different oxidation states shown above, they also have a habit of being produced together in the same reaction and not all are thermally stable.

Q3. Which two oxides show nitrogen in the same oxidation state?
Q4. State what is happening in the reaction shown in eqn 3.4 and what you might expect to see.

$$3NO_{(g)} \rightarrow N_2O_{(g)} + NO_{2(g)} \qquad (3.4)$$

The oxides of nitrogen shown above are, of course, all *different compounds* of nitrogen and oxygen and so the law is not violated but, all the same, sorting out all the different oxides of nitrogen must have been quite a task requiring some inventive thinking, especially when more than one oxide can be made simultaneously in the same reaction.

3.5 PRODUCING NITRIC OXIDE (NO)

The reaction of copper metal with a 50 : 50 mixture of concentrated nitric acid and water produces a mixture of NO and NO_2:

$$Cu \rightarrow Cu^{2+} + 2e^- \qquad (3.5)$$

$$2H^+ + NO_3^- + e^- \rightarrow NO_{2(g)} + H_2O \qquad (3.6)$$

$$4H^+ + NO_3^- + 3e^- \rightarrow NO_{(g)} + 2H_2 \qquad (3.7)$$

Q5. Write balanced equations for the reaction of copper metal with nitric acid to produce:

a) NO.
b) NO_2.
c) Given that NO_2 is very soluble in water, reacting to produce nitric acid, and NO is only slightly soluble, how might you collect a sample of NO?

3.6 GAME FOR A LAUGH?

Joseph Priestly is generally credited as the discoverer of N_2O and, no doubt, recognized some of its intoxicating qualities, which eventually resulted in it getting the name laughing gas. However, as a strict Calvinist, known for his fire and brimstone sermons, he probably considered its recreational use inappropriate. Humphrey Davy had no such qualms and wasn't averse to combining wine and nitrous oxide in an evening's entertainment, occasionally with Messrs Coleridge and Southey. Today, it is still used for mild pain relief during childbirth. The emphasis here, according to my wife, being mild!

3.7 HE WHO LAUGHS LAST...

According to the United States Environmental Protection Agency (EPA),[1] in 2010 "...nitrous oxide accounted for 4% of all greenhouse gas emissions from human activity", with over two thirds of this coming from agricultural soil management. The soil bacteria *Nitrosospira* are responsible for converting ammonium nitrate in the soil into nitrous oxide. The reaction can be summarized in eqn 3.8:

$$NH_4NO_3 \rightarrow N_2O + 2H_2O \qquad (3.8)$$

Q6. The reaction in eqn 3.8 is known as a **comproportionation** reaction. What do you think this means?

Q7. Suggest how the use of ^{15}N isotopes might be used to determine whether *Nitrosospira* are aerobic or anaerobic bacteria.

Nitrous oxide has been estimated to have something like a 300-fold greater impact on the greenhouse effect than carbon dioxide (weight for weight). It might be nice to think that concentrations of nitrous oxide may one day be high enough to have us all rolling around with laughter but, sadly, global temperatures would have become uninhabitable for humans well before then.

3.8 AIR BAGS

Most modern cars are fitted with air bags as a safety feature designed to cushion a rapid impact with your steering wheel, should you be unlucky enough to be in an accident. It goes without saying that the inflation of the air bag has to be rapid and, for this, we rely on sodium azide.

$$2NaN_{3(s)} \rightarrow 2Na_{(s)} + 3N_{2(g)} \qquad (3.9)$$

Q8. Calculate the mass of sodium azide needed to completely inflate a 15 dm^3 air bag. State any assumptions made in your calculation.

3.9 AMINO ACIDS

As the name suggests, amino acids are molecules with both the amine and the carboxylic acid functional group. There are practically an infinite number of molecules that could be called amino acids but life (at least here on Earth) has decided to work with twenty, as far as making proteins is concerned. One such amino acid is shown is Figure 3.1.

Thankfully, this molecule is generally known as arginine (Arg) or even just **R** using the single letter amino acid code.

Q9. Give the empirical **and** molecular formula for arginine.
Q10. Calculate the percentage mass of nitrogen in the molecule.

A protonated arginine molecule is shown in Figure 3.2.

Q11. The pK_a value of the species shown in Figure 3.2 is 12.5. Explain why you could expect arginine to be protonated at a physiological pH of 7.4.

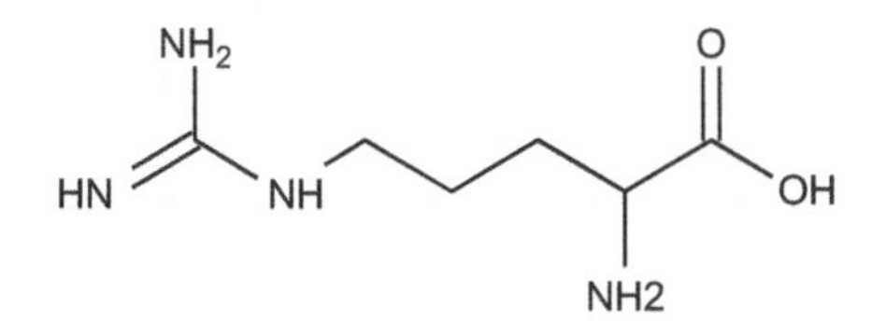

Figure 3.1 Structure of 2-amino-5-carbamimidamidopentanoic acid.

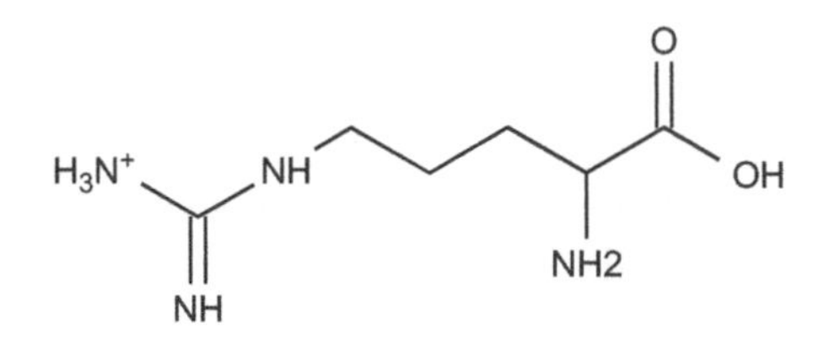

Figure 3.2 Structure of a protonated arginine molecule.

3.10 NITROGEN: STABLE ELEMENT, REACTIVE COMPOUNDS

At the start of this chapter we noted the stability of nitrogen gas in the air and its reactivity in certain compounds. Let's look further into this with nitroglycerin. The molecular formula of nitroglycerin is $C_3H_5N_3O_9$. For comparison let's consider propane fuel and its combustion (eqn 3.10):

$$C_3H_8 + 5O_2 \rightarrow 3CO_2 + 4H_2O \quad \Delta H_c = -2220 \text{ kJ mol}^{-1} \tag{3.10}$$

When propane burns it releases a lot of energy, but the combustion requires oxygen to mix with the propane molecules to ensure a complete combustion. If you want to see incomplete combustion, just close the air hole on a Bunsen burner. You will see the colour of the flame change from blue to orange.

Now look at eqn 3.11, which shows the combustion of nitroglycerin:

$$4C_3H_5N_3O_9 \rightarrow 6N_2 + 12CO + 10H_2O + 7O_2 \tag{3.11}$$

The first thing to note from the equation is that **no external supply of oxygen is required.** Nitroglycerin has more than enough oxygen to self-combust. This kind of self-destructive process is known as a **detonation** and it happens *very* fast. Note also that 4 moles of nitroglycerin (an oily liquid) detonate to produce **35 moles** of gaseous products. In addition, the amount of heat energy released is considerable. Look back at Table 3.1.

Q12. How many kJ of heat energy are released when 6 moles of N_2 are formed from 12 moles of gaseous nitrogen atoms?

A more stable alternative to nitroglycerine is known as TNT (Figure 3.3).

Q13. Give the IUPAC name for TNT

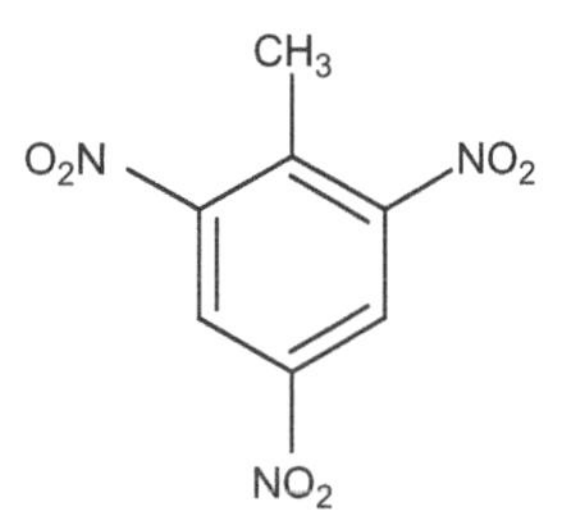

Figure 3.3 Structure of TNT.

So, in fact there is no contradiction between the stability of N_2 and the reactivity of many nitrogen-containing compounds (especially those containing nitrates). The large amount of energy **released** when two atoms of nitrogen combine to form N≡N is the same amount of energy **required** to break the triple bond.

3.11 MOLECULE OF THE 20TH CENTURY?

There are, arguably, two leading candidates for *the* molecule of the 20th century and both contain nitrogen. DNA, the structure of which was finally elucidated in 1953, proved to be the death knell for any lingering Vitalist theory. The hereditary principle turns out to be a chemical polymer of no more than four DNA bases: adenine, cytosine, guanine and thymine. Since that time, genes have been transferred between organisms to alter phenotypes. However, I suspect that the real DNA revolution will take place in the 21st century. I would argue that it is ammonia that is the strongest contender for molecule of the 20th century, and we can't consider ammonia without considering the rise and fall of Fritz Haber.

Explosives require nitrates and, at the turn of the 20th century, there was a shortage. The only large-scale supplies came from deposits in South America, where Peru, Bolivia and Chile had fought a five year war for control of the resource. At the start of World War I, Germany needed a supply of nitrates and it was Fritz Haber who was charged with the task of securing a supply. Haber succeeded in converting nitrogen and hydrogen into ammonia using a range of different catalysts at high pressure.

$$N_2 + 3H_2 \rightleftharpoons 2NH_3 \tag{3.12}$$

Q14. At 450 °C the K_p for the above reaction is 1.45×10^{-3} atm^{-2}. If the equilibrium partial pressures of N_2 and H_2 are 0.432 atm and 0.928 atm, respectively, calculate:

a) the partial pressure of ammonia at equilibrium.
b) what percentage of the gas mixture is ammonia under these conditions.

Having achieved ammonia synthesis, a subsequent oxidation reaction could convert ammonia into nitric oxide, then to nitrogen dioxide and eventually nitric acid (eqn 3.13–3.15).

$$4NH_3 + 5O_2 \rightarrow 4NO + 6H_2O \tag{3.13}$$

$$2NO + O_2 \rightarrow 2NO_2 \qquad (3.14)$$

$$4NO_2 + O_2 + 2H_2O \rightarrow 4HNO_3 \qquad (3.15)$$

Q15. Identify the oxidation state of nitrogen in all of the reactants and products in eqn 3.13–3.15.

Fritz Haber received the 1918 Nobel prize in Chemistry, but Germany's defeat in World War I must have hit the proud nationalist hard. Sadly, worse was to come. With the rise of the National Socialists in the 1930s in Germany, he was forced out of his university post. Despite serving Germany with distinction during World War I, his Jewish ancestry was a matter of record and undeniable. He died a broken man in 1934 in Basel, Switzerland.

3.12 CONCLUSIONS

I said at the beginning of this chapter that nitrogen is a paradoxical element. Left alone it is happy to surround us inert to all but the most energetic chemistry. Forced in to service by mankind it has been responsible for killing millions, and yet also feeding millions. Like its group V neighbor, phosphorus, it is both absolutely essential and utterly devastating.

REFERENCE

1. United States Environmental Protection Agency, *National Greenhouse Gas Emissions Data*, 2013. Available online at http://www.epa.gov/climatechange/ghgemissions/usinventoryreport.html [last accessed 3rd December 2013].

Answers to questions in this chapter can be found on pages 174–176.

CHAPTER 4

Sulfur

> *"And the devil, who deceived them, was thrown into the lake of brimstone, where the beast and the false prophet had been thrown. They will be tormented day and night for ever and ever."*
>
> Revelations 2:10

There are a number of references to "brimstone" in the Bible variously alluding to the Lord's wrath or Hell fire itself. The "burning stone" from which the word is derived is in fact the element sulfur. Large deposits of elemental sulfur can be found at the mouths of volcanoes and, thus, the association with Hell is perhaps understandable, but—and you're probably ahead of me by now—sulfur is an essential element to life. In this chapter we will see that sulfur may have played a key role in life getting started on Earth and it still performs an essential role in protein structures. As an ingredient of gunpowder, it had a major impact on civilization, and in the 19th century Justus von Liebig even suggested that the sulfuric acid production per capita of a nation state could act as a measure of how "developed" a nation was. But to start, let's consider the role of sulfur in alchemy.

4.1 SULFUR AND ALCHEMY

The alchemists weren't really interested in the elements as we would now recognize them. In fact the idea of the element as a fundamental

Around the World in 18 Elements
By David A. Scott

Published by the Royal Society of Chemistry, www.rsc.org

Figure 4.1 An alchemical symbol for sulfur.

and irreducible substance had to wait until Robert Boyle (1627–1691). All philosophical speculation on the nature of matter still had to defer to Aristotle and the fundamental "elements" were considered to be Aether, Fire, Earth, Water and Air. Nevertheless, some substances are naturally attractive to the experimental scientist and Paracelsus (1493–1541) was particularly fascinated with sulfur, mercury and salt, and went so far as to suggest that these substances *"formed everything that was in the four elements"*. Sulfur, with its yellow colour, was fascinating to the alchemists because there seemed to be an intuitive sense that, if it were combined with mercury (with its liquid silver appearance), then maybe gold could be the result. Alas no, but in the modern periodic table you will find the elements sulfur (S) and mercury (Hg), so in a sense, Paracelsus got the ball rolling (Figure 4.1).

4.2 THE SULFUR CYCLE

The Earth's sulfur cycle is complex! There are both biotic and abiotic factors to consider, coupled with the fact that sulfur can adopt a range of oxidation states and, as a result, physical states. Gases include the inorganics, SO_2, SO_3 and H_2S, as well as organics, such as $(CH_3)_2S$ and $(CH_3)_2SO$. Deposits of elemental sulfur occur frequently at areas of geological activity, and iron sulfide (FeS) and iron pyrites (FeS_2) are common but by no means the only ores (see below). In addition the sulfate ion is present in solution in the oceans and rivers and evaporates, such as $CaSO_4 \cdot 2H_2O$.

Q1. Assign an oxidation number to sulfur in each of the following compounds: SO_2, SO_3, H_2S, $(CH_3)_2S$, $(CH_3)_2SO$, FeS, FeS_2 and $CaSO_4 \cdot 2H_2O$.

There is little doubt that human influence (the anthropogenic effect) on the sulfur cycle is considerable, mostly as a result of the combustion of fossil fuels. One estimate in 2000[1] measured global anthropogenic SO_2 output at 1.06×10^8 tonnes per year. With the world's population practically doubling since then, we could conservatively double that quantity to get an up-to-date estimate!

Q2. Calculate the annual volume of SO_2 produced **in km^3** using the 2000 data and assuming 1 mole of SO_2 occupies 24 dm^3 at 298 K.

(When you've finished your calculation, you might like to compare your result with the surface area of Wales, which is 20 779 km^2).

4.3 SULFUR IN THE EARTH'S CRUST

Sulfur is, coincidently, both element number 16 in the periodic table and the 16th most abundant element in the Earth's crust. It can be found as elemental deposits, chemically combined as metal sulfides and hydrogen sulfide, and as sulfates, such as gypsum.

Some of the main sulfide ores are shown in Table 4.1.

Q3.

a) Calculate the percentage mass of sulfur to the nearest whole percentage point in each compound shown in Table 4.1.
b) What other factors would need to be considered in choosing which ore to mine?

Another source of sulfur is hydrogen sulfide gas, which frequently occurs with natural gas deposits. The hydrogen sulfide is removed by absorption in an organic solvent followed by the two reactions shown in eqn 4.1 and 4.2.

$$2H_2S + 3O_2 \rightarrow 2SO_2 + 2H_2O \tag{4.1}$$

$$\text{then } 2H_2S + SO_2 \rightarrow 3S + 2H_2O \tag{4.2}$$

Table 4.1 The main sulfide ores.

Ore	*Formula of main compound*
Molybdenite	MoS_2
Pyrite	FeS_2
Sphalerite	ZnS
Galena	PbS
Chalcopyrite	$CuFeS_2$

Q4.

a) Give the oxidation number of sulfur in all of the reactants and products in the eqn 4.1 and 4.2.
b) Combine eqn 4.1 and 4.2 and calculate the mass of sulfur you might expect to collect from 500 dm^3 of H_2S. (Volumes measured at room temperature and assume 100% conversion.)

Sulfur is perhaps more regularly encountered in it higher oxidation state in compounds, such as sulfuric acid or as sulfates. Let's start by looking at the world's most important acid.

4.4 SULFURIC ACID

The manufacture of sulfuric acid is big business. In the year 2000 the worldwide production was in the region of 160 million tonnes.[2] It is used as an acid, an oxidizing agent, a dehydrating agent, drying agent and a catalyst, and its production must be as economically viable as possible. Industrially, it is made by the contact process and the key step in the reaction, shown in eqn 4.3, is carried out at 400 °C with a vanadium V oxide catalyst.

$$2SO_2 + O_2 \rightleftharpoons 2SO_3 \qquad \Delta H = -197 \text{ kJ mol}^{-1} \tag{4.3}$$

Q5.

a) Use the standard entropy values in Table 4.2 to calculate the entropy change for the reaction in eqn 4.3 at 25 °C.
b) Now calculate the total entropy change for the system and surroundings at 25 °C.
c) Suggest why the reaction is carried out at 400 °C in industry.
d) Using $\Delta S_{total} = R\ln K$, calculate a value for K_p at 400 °C.

The K_p value you've calculated above hopefully shows you that the reaction is thermodynamically feasible under the given conditions. However, this tells us nothing about the **rate** at which this K_p value is

Table 4.2 Entropy values.

Substance	*Standard entropy/J K^{-1} mol^{-1}*
O_2	102.5
SO_2	248.1
SO_3	256.8

attained and, without a catalyst, the reaction is uneconomically slow. How then does vanadium V oxide catalyze this reaction? The answer is shown below in eqn 4.4 and 4.5:

$$SO_2 + V_2O_5 \rightarrow SO_3 + 2VO_2 \tag{4.4}$$

$$2VO_2 + \tfrac{1}{2}O_2 \rightarrow V_2O_5 \tag{4.5}$$

Q5.

e) Combine eqn 4.4 and 4.5 to give the overall equation for the contact process.
f) Explain in terms of redox what is happening to vanadium in this process.

4.5 SULFATES

Sulfates, as the name suggests, implies sulfur combined with oxygen in its various forms. There are quite a few varieties!

SO_3^{2-} (sulfate IV or sulfite), SO_4^{2-} (sulfate VI or just sulfate), $S_2O_3^{2-}$ (thiosulfate), $S_4O_6^{2-}$ (tetrathionate), $S_2O_8^{2-}$ (peroxodisulfate) and others. Despite the best endeavours of the IUPAC, sulfate IV is still called sulfite and the name sulfate rarely includes the VI in brackets. Old habits die hard it would seem! Here after, I will use the "old" names. We'll start by looking at the sulfates.

Q6. The sulfur atom in the sulfate ion uses the 3s, 3p and two of the 3d orbitals for bonding. Draw "electrons in boxes" to show how this explains sulfur's +6 oxidation state.

The shape of the sulfate ion, however, is tetrahedral and one suggestion for its shape is sp^3 hybridization, with the two 3d orbitals overlapping to give two π-type bonds with two oxygen p orbitals. Thus, sulfate is usually represented as shown in Figure 4.2.

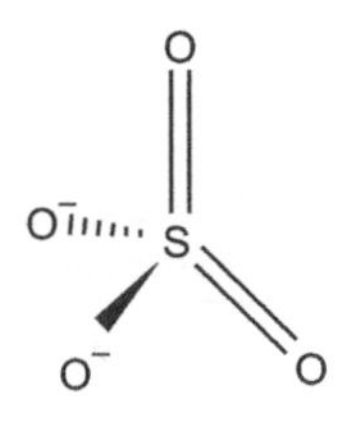

Figure 4.2 Structure of a sulfate ion.

However, even this "picture" needs to be modified as the sulfate anion has no discrete double and single bonds and is better understood as a resonance stabilized structure in which all S–O bond lengths are the same, at 0.149 nm (intermediate between 0.145 nm for an average S=O and 0.157 nm for S–O). The −2 charge is delocalized over the whole anion. As a result, the sulfate anion has the geometry of a tetrahedral.

Over the years I have been frequently struck with the inter-relationship between symmetry and stability in chemistry and, thus, I was pleased when I came across the following passage in Primo Levi's *The Periodic Table*.

> *"It makes you think of something solid, stable, well linked. In fact it also happens in chemistry as in architecture that "beautiful" edifices, that is symmetrical and simple are also the most sturdy: in short, the same thing happens with molecules as with the cupolas of cathedrals or the arches of bridges."*[3]

In the passage above, Levi talks about the molecule alloxan (2,4,5,6-pyrimidinetetrone) but the observation holds equally well for the sulfate anion; molecules and molecular ions seem to "like" symmetry!

The resonance stability of the sulfate anion helps explain its stability and why it is such a weak conjugate base in aqueous solution.

Q7.

a) With reference to eqn 4.6, explain what is meant by the term conjugate base.

$$HSO_4^{-}{}_{(aq)} \rightleftharpoons H^{+}{}_{(aq)} + SO_4^{2-}{}_{(aq)} \quad (4.6)$$

b) If the pK_a value for the reaction above is 2.0, write out an expression for SO_4^{2-} acting as a base and calculate a K_b value for the reaction giving appropriate units.

Most sulfates are soluble to a greater or lesser degree but one important exception is **barium sulfate**. Because of its extremely low solubility, the sulfate in solution can be quantified using a technique called gravimetric analysis. In this case soluble sulfate ions are precipitated out as barium sulfate and then the insoluble salt is dried and weighed.

Q8.

a) An analysis to determine the percentage mass of $CaSO_4 \cdot 2H_2O$ in an impure sample of gypsum gave the following results:

Initial mass of sample = 10.30 g, crushed and dissolved in 250 cm^3 of solution.

Mass of $BaSO_4$ precipitated from 25 cm^3 aliquot = 1.21 g.

Given this information, what is the percentage of $CaSO_4 \cdot 2H_2O$ in the impure sample? Suggest why the storage conditions of the sample may affect your results.

b) The solubility product (K_{sp}) for barium sulfate = $[Ba^{2+}][SO_4^{2-}] = 1 \times 10^{-10}$ mol^2 dm^{-6} at 25 °C.

Calculate the maximum mass of $BaSO_4$ that could dissolve in 50 dm^3 of water at 25 °C. Barium ions are toxic and yet barium sulfate is given as a barium meal to "gastric" patients prior to X-ray analysis. Can you explain this apparent contradiction?

Q9.

a) Calcium hydrogen sulfite, $Ca(HSO_3)_2$ (sometimes called calcium bisulphate), is a mild reducing agent in aqueous solution and is used as a food preservative (E227).

Give the oxidation number of sulfur in this compound and write equations to show how calcium hydrogen sulfite forms an acidic solution.

b) The hydrogen sulfite anion has a $pK_a = 7.2$. Calculate the pH of a 0.01 mol dm^{-3} solution of $Ca(HSO_3)_2$.

4.6 SODIUM THIOSULFATE: A USEFUL ANALYTE

Sodium thiosulfate has a number of things to recommend itself to the analytical chemist: it is a relatively stable compound in air, with a molecular mass of 158 g mol^{-1} and it is very soluble in water.

Q10.

a) Why is the comparatively large relative formula mass and high solubility of sodium thiosulfate an advantage when making up a standard solution?

b) Use the data in Table 4.3 to show how a solution of sodium thiosulfate reduces iodine to iodide and calculate a standard electrode potential for the reaction.

Table 4.3 Standard electrode potentials for the reduction of iodine to iodide and tetrathionate to thiosulphate.

Half cell	*Standard electrode potential E^{θ}/V*
$I_2 + 2e \rightarrow 2I^-$	+0.54
$S_4O_6^{2-} + 2e \rightarrow 2S_2O_3^{2-}$	+0.09

c) What volume of 2×10^{-3} M $Na_2S_2O_3$ would be required to oxidize 50 cm^3 of a 0.003 M solution of iodine and what indicator would be required to determine the end point of the titration?

4.7 ORGANIC SULFUR

As unlikely as it seems, it was a compound called an azo dye that led to a class of anti-bacterial compounds called sulfonamides. Using dyes to stain cells so that they could be more easily seen under microscopes was a well-developed technique and, thus, it was probably a lucky accident that showed some of these stains had anti-bacterial properties.

In his book, *Life Saving Drugs*,[4] John Mann recalls the famous success of sulfapyridine (Figure 4.3):

> *"It [sulfapyridine] achieved star status once it was revealed that it had been used to save the life of Winston Churchill when he contracted pneumonia during a visit to North Africa in December 1943."*

However one of the problems of sulfapyridine is its pH-dependent solubility.

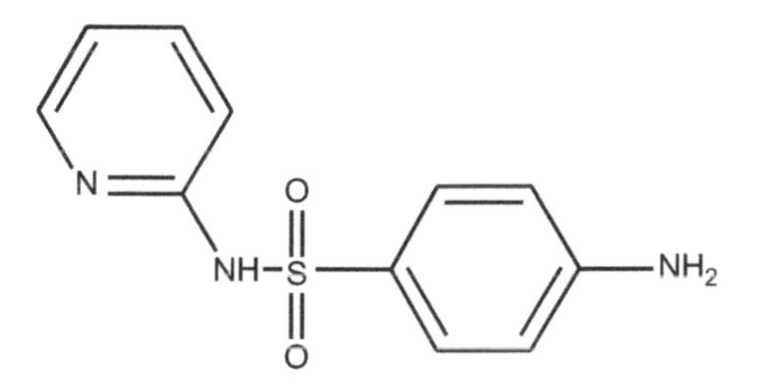

Figure 4.3 Structure of sulfapyridine.

Q11. With reference to the structure in Figure 4.3, suggest which pH conditions are likely to favour the solubility of sulfapyridine in aqueous solution.

Q12. One of the analytical techniques that can be used to help identify organic molecules is electron impact (EI) mass spectrometry. Analysis of sulfapyridine using this technique gives a significant peak at $m/z = 92$ but only the slightest trace at $m/z = 249$. Outline the basic principles behind EI mass spectrometry and explain the m/z observations.

4.8 SULFUR IN PROTEINS

Of the twenty amino acids that constitute the starting monomers of all eukaryotic proteins, two contain sulfur: cysteine and methionine. Cysteine molecules play a vital role (quite literally) in stabilizing the tertiary structure of proteins.In an enzyme-mediated redox reaction cysteine residues can form a covalent bond as shown in Figure 4.4.

Q13.

a) Explain why this reaction is classified as a redox reaction and give a value for the C–S–S bond angle explaining your answer.
b) Why might you expect the S–S bond to be more susceptible to electrophiles than the C–C bond?

Human hair has a high sulfur content (in the region of 5% by mass). *Tricothiodystrophy* (or TTD), however, is a genetic condition that causes brittle hair due to sulfur deficiency. In many cases this medical condition is associated with other symptoms, which may suggest a more general problem with the metabolism of cysteine and/or methionine.

4.9 SULFUR: A ROLE IN GETTING LIFE STARTED?

One of the most conserved features of any living system is the ability to generate energy from a proton gradient. ATP is synthesized in

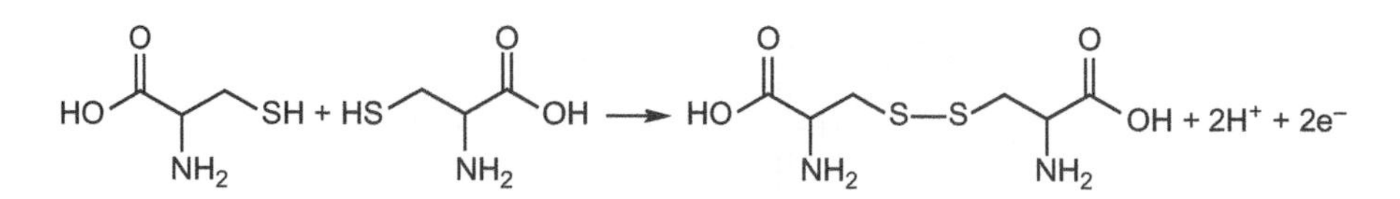

Figure 4.4 Enzyme-mediated redox reaction forms a disulfide bond between two cysteine molecules.

mitochondria by coupling it to a trans-membrane proton potential. It now seems that long before the emergence of photosynthesis, in which protons are stripped from water with the release of oxygen, hydrogen sulfide was being oxidized to generate the source of protons resulting in the precipitation of sulfur:

$$H_2S \rightarrow S + 2H^+ + 2e^- \quad (4.7)$$

In 1880 the Russian microbiologist Sergei Winogradsky (1856–1953) observed bacteria from the genus *Beggiatoa* depositing granules of sulfur internally when "fed" on hydrogen sulfide, showing that this particular genus was capable of getting its energy from the oxidation of sulfide. Such organisms are referred to as Chemotrophs. *Thiobacillus thiooxidans* can go even further and oxidizes hydrogen sulfide to sulfuric acid.

Q14. Write a balanced half equation for the oxidation of hydrogen sulfide to sulfate ions.

Recent research seems to show that bacteria and archaea are not alone in being able to utilize hydrogen sulfide. In a paper by O'Brien and Vetter[5] mitochondria from the organism *Solemya reidi* (a species of clam) were shown to generate protons by oxidizing hydrogen sulfide to thiosulfate. In order to follow the metabolism of sulfur the organism was supplied with a supply of Na_2S, which was enriched with an ^{35}S isotope. ^{35}S decays by β emission.

Q15.

a) Write a balanced equation for the decay of the ^{35}S isotope and identify the element produced.

All radioactive decay has first order kinetics and, thus, the relationship $\ln c - \ln c_0 = -kt$ holds, where c is the concentration of the isotope at time t, c_0 is the initial concentration of the isotope and k is the decay constant for the isotope. The half-life of ^{35}S is 44.3 days.

Q15.

b) Calculate the time it takes for 10 mg of isotope to decay to 1 mg.

4.10 SULFUR AND HYDROTHERMAL VENTS

There was a time when it was thought that all life on Earth was dependent, either directly (plants and some prokaryotes) or indirectly (animals and other prokaryotes), on photosynthesis. The discovery of prokaryotes living in the depths of the oceans called for a re-think. Many miles below the ocean surface it became clear that such organisms derived their energy sources from non-photosynthetic sources and many scientists now believe that such Chemotrophs may have been the first living systems on planet Earth. Some evidence puts the appearance of life on Earth as early as 3.8 billion years ago, at a time when life at the Earth's surface would have been too unstable to survive.

4.11 CONCLUSIONS

In some ways, sulfur is a bit like phosphorus. It is a moderately reactive non-metal, which is essential to life. Phosphorus, however, occurs almost exclusively as phosphate V and, because it doesn't form a gaseous oxide, it has a relatively straight forward biogeochemical cycle. Sulfur, which exits commonly in five oxidation states and in all three phases, provides scientists with more of a challenge!

REFERENCES

1. Z. Klimont, S. J. Smith and J. Cofala, The last decade of global anthropogenic sulfur dioxide: 2000–2011 emissions, *Environ. Res. Lett.*, 2013, **8**, 014003.
2. P. Levi, *The Periodic Table*, 1986, Abacus, London, 120.
3. Royal Society of Chemistry, Welcome to Alchemy? 2007. Available online at: http://www.rsc.org/Education/Teachers/Resources/Alchemy/index2.htm. Accessed 19th December 2013.
4. J. Mann, *Life Saving Drugs: the Elusive Magic Bullet*, 2004, Royal Society of Chemistry, Cambridge.
5. J. O'Brien and R. D. Vetter, *J. Exp. Biol.*, 1990, **149**, 133–148.

Answers to questions in this chapter can be found on pages 177–180.

CHAPTER 5

Calcium

Of all of the elements, it is calcium that I have the greatest difficulty in convincing younger pupils that it is **a metal**. As is often the case, irrefutable logic gets in the way. The reasoning can be broken down into the following syllogism:

All things that contain metals have typical metallic properties, e.g., shiny, solid, malleable, etc.

Milk, which we all know is rich in calcium, has none of these properties.

Therefore, calcium is not a metal.

I'm not suggesting for a minute that this reasoning process is deliberate and conscious but, over the years, I've become convinced that this, or something very similar, is happening in many cases. The logical reasoning is flawless and thus the conclusion is valid but not, of course, actually TRUE. The reason is because our initial premise is wrong! All things that contain metals do not have typical metallic properties because, when metals react, they lose their individual properties. I start with this rather pedantic analysis because, as the developmental biologist Lewis Wolpert has pointed out, scientific reasoning and intuitive common sense can frequently be contradictory and thus, not surprisingly, students can feel a bit confused.

In this chapter we will encounter calcium's role in providing evidence for the Big Bang. The importance of calcium compounds in

Around the World in 18 Elements
By David A. Scott

Published by the Royal Society of Chemistry, www.rsc.org

structures from the smallest coccoliths (marine organisms) to the largest buildings, its abundance in hard water and its role in cellular development and communication. In none of these examples is calcium showing the typical physical properties that are commonly associated with metals.

5.1 DALTON'S SYMBOL FOR LIME

It wasn't until 1808 that Humphrey Davy (1778–1829) finally separated calcium from its oxide using the newly discovered method of electrolysis. The oxide of calcium (lime), however, was known by the ancient civilizations and, until Davy's isolation, had been thought to be one of the "earthy" elements as it neither dissolved in water nor decomposed on heating. John Dalton (1776–1844) was one of the first to use a symbolic notation for the chemical elements and his chosen symbol for lime can be seen in Figure 5.1.

Today, we write lime as CaO and we can get an idea of its stability by calculating its **lattice enthalpy** using a **Born–Haber cycle**.

Q1.

a) Define the term lattice enthalpy.
b) Draw a Born–Haber cycle for the formation of calcium oxide and use the data in Table 5.1 to calculate a value for the lattice enthalpy of calcium oxide.
c) Explain why the second electron affinity for oxygen is endothermic.

Figure 5.1 John Dalton's chemical symbol for lime (CaO).

Table 5.1 Enthalpy values for the formation of calcium oxide.

Process	*Enthalpy/kJ mol^{-1}*
Atomization of Ca	193
First ionization energy of Ca	590
Second ionization energy of Ca	1150
Bond enthalpy of O_2	496
First electron affinity for O	−142
Second electron affinity for O	844
Enthalpy of formation CaO	−653

The theoretical values for lattice enthalpies can be calculated using the Born–Lande equation. Without going into this equation in too much detail, we can note here that the force of attraction between the two ions is proportional to the product of the charges on the two ions divided by the sum of the ionic radii. The assumption made here is that the ions are perfect spheres and thus the electrostatic charge is equally distributed over the ionz.

(d) Compare the theoretical lattice enthalpy of CaO with its experimental value shown in Table 5.2. Suggest why there is a difference between the two values.

5.2 SLAKED LIME

Dalton's symbol for hydrated or slaked lime is shown in Figure 5.2. He represented slaked lime as lime + water. Dalton believed water to be made from hydrogen and oxygen in the ratio of 1 : 1.† Although we now know this to be incorrect, we can see in his notation the beginnings of the chemical formula in use today.

Table 5.2 The theoretical and experimental lattice enthalpies for CaO and BaO.

Substance	*Theoretical lattice enthalpy/kJ mol^{-1}*	*Experimentally determined lattice enthalpy (using Hess's law)/kJ mol^{-1}*
CaO	−3477	−3536
BaO	−3042	−3152

Figure 5.2 John's Dalton's chemical symbol for slaked lime.

†The smaller circle with the black dot inside is Dalton's symbol for hydrogen; the other symbol is oxygen.

5.3 EXTRACTION OF CALCIUM

As the fifth most abundant element in the Earth's crust, calcium is not in short supply. However, as we have seen, like other reactive metals, calcium is not easily separated from its compounds. Industrially, calcium is extracted by the electrolysis of its molten chloride.

Q2.

a) Describe the bonding in solid calcium chloride and suggest why its melting point is relatively high, at 772 °C.
b) Write equations for the reduction of calcium ions at the cathode and the oxidation of chloride at the anode during electrolysis.
c) An industrial electrolysis plant uses a current of 15 000 A. What mass of calcium metal, to the nearest kg, could be extracted from its chloride if this current was passed through molten calcium chloride for 8 h? (Faraday constant = 96 480 C.)
d) Distinguish between the terms *hazard* and *risk* and consider both factors in relation to the extraction of calcium by this process on an industrial scale.

5.4 IDENTIFYING CALCIUM FROM ITS ATOMIC EMISSION SPECTRUM

Like all elements, calcium can be identified from is atomic emission spectrum. Each element has a unique emission spectrum that acts as a "fingerprint" for that element. When excited by energy, electrons are promoted from lower to higher energy levels and many then return to lower energy levels emitting a specific frequency of light energy. Some elements emit light in the visible part of the spectrum and the colour observed is an amalgamation of all of the quantized line frequencies emitted. One particular frequency emitted by calcium is at 616 nm. The relationship between the frequency of a photon of light is given by $E = hf$ and the relationship between frequency and wavelength is $c = f\lambda$.

Q3.

a) Calculate the frequency of this wavelength of light giving appropriate units.
b) Calculate the energy of a photon of this wavelength of light giving units.

It just so happens that two particular lines in the spectrum of calcium are useful for measuring the so called "red shift", which enables astronomers to determine how far galaxies are from us and how fast they are moving away. The Virgo cluster, for example, has been estimated at a distance of 78 million light years away and is receding at a speed of 1 200 000 m s^{-1} (that is 0.4% of the speed of light)!

5.5 CALCIUM, CEMENT AND CONCRETE

Back down to Earth. After water, concrete is the most extensively used material by humans. Today, its manufacture is also responsible for 5% of anthropogenic CO_2 output. Its wide-scale use in construction possibly dates back to the earlier Egyptian dynasties and, in Roman times, Pliny the Elder talks of a mixture of one part lime to four parts sand to make mortar. Some of the largest man-made constructions on Earth, such as the Hoover Dam and the recently constructed Three Gorges Dam on the Yangtze in China, are constructed from concrete.

Concrete is essentially Portland cement + water + aggregate. A typical Portland cement has the following composition by mass:

70% calcium oxide, CaO
20% silicon dioxide, SiO_2
5% aluminium oxide, Al_2O_3
3% iron III oxide, Fe_2O_3
2% hydrated calcium sulphate, $CaSO_4 \cdot 2H_2O$

Q4. Calculate the total mass of calcium in 1 tonne of Portland cement.

One of the key reactions in the formation of concrete is shown in the following reaction (eqn 5.1):

$$Ca_3SiO_5 + H_2O \rightarrow Ca_2SiO_4 + Ca(OH)_2 \quad \Delta H = -87 \text{ kJ mol}^{-1} \qquad (5.1)$$

Once the hydrated Ca_2SiO_4 begins to cool, it crystallizes and binds the aggregate to make concrete. Similar reactions occur with the other components of Portland cement and the properties of the final product can be modified by altering the original composition of the cement.

Q5. The following experiment was carried out to obtain an enthalpy value for the reaction shown in eqn 5.1:

- 14.8 g of Ca_3SiO_5 was placed into a copper calorimeter of mass 50.0 g.

Table 5.3 Specific heat capacities for water and copper.

	Water	*Copper*
Specific heat capacities/J g^{-1} °C^{-1}	4.2	0.4

- The copper calorimeter was then placed in an insulated beaker containing 100 g of water.
- The equipment was left for a period of about 10 min.
- The initial temperature of the water was recorded at 16.5 °C.
- 2.0 g of water was added to the copper calorimeter and the mixture was stirred.
- The maximum temperature of the water was recorded at 32.5 °C.

a) Use the information above to calculate a value for the enthalpy of reaction. You will need the data in Table 5.3.
b) How does your experimental value compare to the one shown in eqn 5.1? Suggest reasons for any differences.

5.6 HARD WATER, SCUM AND SCALE

If you live in an area of hard water you will be all too familiar with scum and scale. Rainwater percolating through chalk and limestone rocks will pick up calcium and carbonate ions, and eventually the concentration of calcium hydrogen carbonate (also known as calcium bicarbonate) will build up due to excess carbon dioxide from the atmosphere, as shown in eqn 5.2.

$$CaCO_{3(aq)} + H_2O_{(l)} + CO_{2(g)} \rightleftharpoons Ca(HCO_3)_{2(aq)} \quad (5.2)$$

This is fine if the water stays cold but the problems start when it is heated, as it is in a kettle or washing machine. The following decomposition reaction (eqn 5.3) takes place:

$$Ca(HCO_3)_{2(aq)} \rightarrow CaCO_{3(s)} + H_2O_{(l)} + CO_{2(g)} \quad (5.3)$$

Q6. A solution of calcium hydrogen carbonate contains 160 g calcium hydrogen carbonate per dm^3 of solution. What mass of calcium carbonate could be formed if all of the calcium hydrogen carbonate in 5 dm^3 of this solution completely decomposes?

5.6.1 Scum

Sadly, even cold hard water has its issues. The molecule shown in Figure 5.3 is sodium stearate, which is a molecular species present in many soaps. Sodium stearate is soluble in water and dissociates to

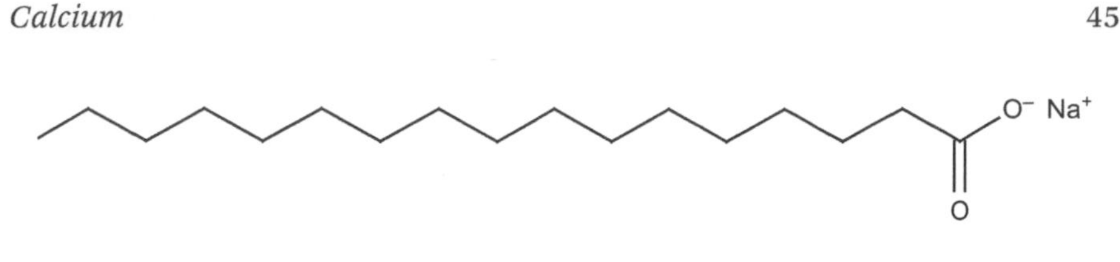

Figure 5.3 Structure of sodium stearate.

form sodium ions and stearate ions. However, in hard water the high concentration of calcium ions causes calcium stearate to precipitate out as scum.

Q7.

a) Explain what is meant by the terms *hydrophobic* and *hydrophilic* and identify which parts of the stearate molecule the terms apply to.
b) Calculate the molecular mass of a sodium stearate molecule.

5.6.2 How Hard is Your Water?

Calcium ions are the main cause of hardness in water, together with its group II neighbour, magnesium. Other species that can contribute to water hardness are iron III and aluminium but these ions are usually present in such low concentrations as to make a negligible contribution. Hardness of water can be measured by a complexiometric titration method using ethylenediaminetetraacetic acid or EDTA for short. In aqueous solution the EDTA is deprotonated to form the $EDTA^{4-}$ anion, which can then form a stable complex with calcium and magnesium ions in solution (Figure 5.4).

5.6.3 Quantifying Calcium Ions in Solution

The water sample to be analyzed is pipetted into a conical flask and a few drops of Eriochrome® Black T (EBT) are added, which immediately forms a complex with calcium and magnesium ions, forming a red colour. As a solution of EDTA is titrated into the sample, it displaces the Eriochrome® Black T to form a very stable hexadentate complex (see Figure 5.4). The end point of the reaction occurs when all of the calcium and magnesium ions are removed from the EBT complex and the indicator turns blue.

Q8.

a) A 25 cm^3 sample of water was titrated against a buffered EDTA solution of concentration 4.8×10^{-3} mol dm^{-3}. The

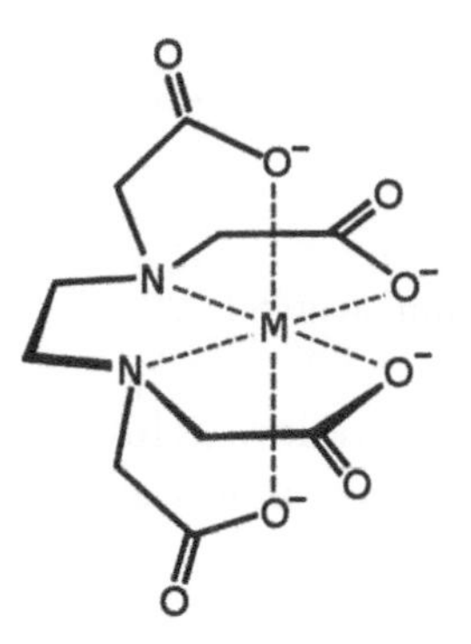

Figure 5.4 Ca^{2+} ion EDTA complex ($M = Ca^{2+}$).

mean end point titre was 15.30 cm^3. Assuming that the hardness was only due to dissolved $CaCO_3$, calculate the concentration of $CaCO_3$ in this solution in mg dm^{-3}.

$$Ca^{2+} + EDTA^{4-} \rightarrow [CaEDTA]^{2-} \quad (5.4)$$

b) Explain what is meant by a hexadentate ligand and describe the geometry of the metal ion–EDTA complex.
c) What kind of bond is formed between the ligands and the metal ions in the complex?

5.7 LIMESTONE AND SOIL pH REGULATION

Soil pH is an important factor for arable farmers and both limestone and slaked lime can be used to adjust soil pH. In many cases fertilizers, such as ammonium nitrate, leave the soil appreciably acidic and this can cause problems as it increasing the solubility of some ions, such as Al^{3+}. Limestone provides a source of carbonate ions, which help to neutralize the acid (eqn 5.5).

$$CaCO_{3(s)} + 2H^+_{(aq)} \rightarrow Ca^{2+}_{(aq)} + CO_{2(g)} + H_2O_{(l)} \quad (5.5)$$

5.8 CALCIUM'S BIOLOGICAL ROLE

Calcium has the largest concentration gradient of any elemental species across the plasma membrane of a living cell. Extracellular fluids have calcium ion concentrations in the order of 1×10^{-3} mol L^{-1} (comparable to seawater), whereas cytoplasmic concentrations are in the order of 1×10^{-7} mol L^{-1}. There are only two explanations for such a gradient: the first is a highly active process in which calcium ions are pumped out of the cell, and the second is the sequestration of calcium ions by specific proteins. Both processes happen.

Q9. If the volume of a bacterial cell is given as 1×10^{-15} dm^3, how many calcium ions would you expect to find in a single bacterial cell and how many would you expect to find in an equal volume of extracellular fluid?

Why does the living cell work so hard to minimize the intracellular concentration of calcium ions? One possible answer concerns the universal unit of energy currency, adenosine triphosphate (ATP). The hydrolysis of ATP releases energy (see Chapter 1) by releasing free phosphate ions.

Q10.

a) The solubility product (K_{sp}) for $Ca_3(PO_4)_2$ at 25 °C $= 2.07 \times 10^{-33}$ mol^5 dm^{-15}. Calculate the maximum mass of $Ca_3(PO_4)_2$ that could dissolve to make a 1 dm^3 solution at this temperature.
b) Given that the concentration of phosphate ions in the cell cytoplasm is reasonably high, suggest why it is necessary to remove free calcium ions from the solution inside the cell.

It is quite possible that the evolution of exoskeletons based on $CaCO_3$, as well as $Ca_3(PO_4)_3$, may have resulted from the precipitation of these compounds on the outer surface of the membrane. The importance of this process is significant in the evolutionary fossil record.

5.8.1 Intracellular Calcium

Intracellular calcium ion concentrations can be regulated by a range of calcium binding proteins. One such protein is called **calsequestrin**, which is the major calcium storage protein in skeletal and cardiac muscle.Calsequestrin has a number of calcium binding sites that enable it to bind calcium ions with varying degrees of affinity. One amino acid involved in binding calcium ions, called aspartic acid, is shown in Figure 5.5.

Q11.

a) Give the IUPAC name for aspartic acid.
b) What feature of aspartic acid enables it to bind to a calcium ion?

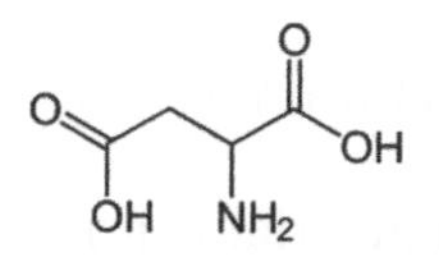

Figure 5.5 Structure of aspartic acid.

c) Suggest why the strength of this bond might be pH dependent.

5.8.2 Calcium is there at the Very Beginning of life...

It has been known for sometime that calcium ions play a key role in intra- and inter-cellular communication and, more recently, it has been shown that when a sperm cell fuses with an egg cell a signal transduction pathway is initiated that causes a "wave" of calcium ions to flow across the egg cell. This reaction is thought to be important in ensuring that multiple cell fertilizations do not take place, although the exact mechanism of this calcium wave is still not completely understood.

5.8.3 ...And at the End

Given calcium's pivotal role at the beginning of life, it is ironic that it is frequently there at the end too! As a strongly basic substance it will aid in the decomposition of a corpse, whilst preventing microbial growth and the resulting odours that can emerge. As it is only slightly soluble it will also be slowly released as the decomposition progresses, remaining active for many years.

5.9 CONCLUSIONS

The majority of exoskeletons are made predominantly from insoluble calcium salts. It is thanks to these minerals that we have a fossil record of some of the earliest living organisms. Indeed, life seems to have stumbled upon bio-mineralization relatively suddenly, giving rise to a moment in history now called the Cambrian Explosion (about 540 million years ago). The sudden appearance of these fossils has suggested to some biologists that evolution takes place in rapid bursts followed by periods of comparative stasis rather than a slow continuous process. It continues to be a topic of fierce debate but one in which calcium chemistry has provided some tantalizing clues.

Answers to question in this chapter can be found on pages 181–183.

CHAPTER 6

Lead

"You that choose not by the view,
Chance as fair and choose as true!
Since this fortune falls to you,
Be content and seek no new..."

So go the opening lines of the verse that greets Bessanio when he opens, not the gold or the silver, but the lead casket in Shakespeare's *The Merchant of Venice*. It seems that for Shakespeare the dull solidity of "base" lead had rather more to recommend it than the more showy "nobles": silver and gold.

Lead was one of the seven metals recognized in the alchemical tradition (the others being gold, silver, mercury, iron, copper and tin). Each of these seven metals was associated with a "planet"† and, in the case of lead, the planet was Saturn. Today, the mystical traditions of alchemy and astrology are incompatible with science but Saturn—slow moving in the heavens—was associated with lead and anyone born "under the influence" of Saturn (in astrological terms) was destined to be gloomy and indolent (hence the word saturnine). Astrologists continue to engage the credulous with "predictions" gleaned from the planetary positions, but there is little doubt that lead exposure has a negative impact on health. It is quite possible that

†The word planet comes from the Greek word wanderer. Today, we would not consider either the Sun or the Moon as a planet but, prior to the heliocentric model of the solar system, both were considered so.

Around the World in 18 Elements
By David A. Scott

Published by the Royal Society of Chemistry, www.rsc.org

early metals workers who were exposed to lead vapour would have developed many symptoms of lead poisoning: insomnia, depression and nausea, which are consistent with "saturnine" personality traits. A recent study[1] showed a correlation (if not necessarily a causal link) between the end of the use of tetraethyl lead (and additive in petrol) and a reduction in violent crime. Today, no-one seriously believes that the position of Saturn at your birth determines your personality but, equally, few would dispute the damaging influence of lead on health. We often perceive early systems of belief as childish and/or ridiculous but there would have been a certain logical consistency to them. It was, after all, the astrologers who were the forerunners of today's astronomers and the alchemists who preceded the chemists.

In this chapter we will start by taking a historical look at lead: was lead the cause of the demise of the Roman Empire as has been suggested? Compounds of lead were there at the beginning of the cosmetics industry, and artists have used some of its more colourful compounds as pigments. More recently, most new cars have been fitted with a lead accumulator battery and lead isotopes have helped us age the Earth, BUT (and lets clear this one up from the off!)...you will NOT find lead in pencils.

6.1 PLUMBERS

In Chapter 5 on calcium we were introduced to the early chemical notation of John Dalton, but it was the Swedish chemist Jacob Berzelius who was largely responsible for the letter notation now used universally by chemists. It is no coincidence that the seven metals that were known to the ancients were given symbols derived from their Latin names and, hence, lead has the symbol Pb from *Plumbum*. Since Roman times, lead has been used for water pipes and, although now largely being phased out, you still call a *plumber* to fix your leaks (if you can find one)!

The symbol shown in Figure 6.1 represents lead and the planet Saturn in the astrological lexicon. The number seven was thought to

♄

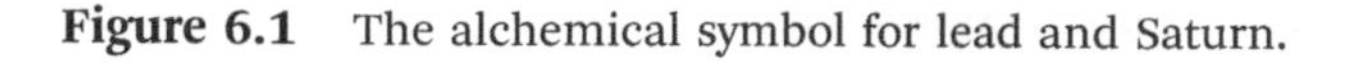

Figure 6.1 The alchemical symbol for lead and Saturn.

have a special, almost mystical significance in "pre-scientific" times: the seven days of Genesis, the seven tones in the diatonic scale are two examples. Even the great Isaac Newton, on splitting white light into the rainbow, "invented" a seventh colour (indigo) to fit with the symbolically perfect seven. Many historians of science were rather horrified to discover that Newton, who seemed to epitomize the ideal of the rational man, spent a great deal of time pursuing the "Philosopher's Stone".

Q1.

a) A sample of lead has the isotopic abundance, ^{204}Pb 1.4%, ^{206}Pb 24.1%, ^{207}Pb 22.1% and ^{208}Pb 52.3%. Calculate the relative atomic mass of lead from this data.
b) A lead atom has an atomic radius of 0.154 nm. Taking the Earth's equatorial diameter to be 12 756 km, how many lead atoms would be needed to circle the Earth's equator?
c) Use your answer to parts a) and b) to calculate the mass of this quantity of lead giving your answer in suitable units.

6.2 LEAD AND SILVER

Silver is commonly found with lead ores and it is likely that lead was initially seen as a by- product from the process of isolating silver. The ancient process known as cupellation is thought to have its origins as far back as the Mycenaean Age (1600 BCE–1100 BCE) and involves separating noble metals (gold and silver) from the base metals, which included copper and zinc, as well as lead. The alloyed mixture of metals was placed in a bone and ash cupel and heated to high temperatures, whereupon the base metals would react with oxygen to form oxides and be absorbed into the lining of the cupel, leaving the unadulterated noble metals. At the end of the reaction, the crust remaining on the inside of the cupel was largely an oxide of lead with a red/orange or yellow colour. In China during the Tang Dynasty the yellow "litharge" (lead II oxide) was used as a cosmetic on women's foreheads as yellow was thought to be an auspicious colour. In western cultures "white lead" was made by the reaction of lead with moist fumes of ethanoic acid. The white compound formed was probably a mixture of lead carbonate ($PbCO_3$) and lead acetate $Pb(CH_3CO_2)_2$ and it was this that gave Elizabeth I her characteristic pallor.

6.3 LEAD EXTRACTION

Today, lead is an important metal in its own right. One of the most important lead ores is galena, from which lead can be extracted in the following reaction (eqn 6.1 and 6.2):

$$2PbS + 3O_2 \rightarrow 2PbO + SO_2 \quad (6.1)$$

$$\text{then} \quad 2PbO + PbS \rightarrow 3Pb + SO_2 \quad (6.2)$$

Q2.

a) Assign oxidation numbers to lead and sulfur in eqn 6.1 and 6.2 and identify which of the two processes is classified as a redox reaction.
b) What is the **overall** atom economy for the process?
c) If 500 kg of lead were extracted from 850 kg of galena, what would be the percentage mass of lead sulphide in the galena sample? (You can assume both reactions give a 100% yield.)

Lead II oxide, once formed, can slowly react with carbon dioxide in the air to form lead carbonate according to eqn 6.3:

$$PbO_{(s)} + CO_{2(g)} \xrightarrow{\Delta H_1} PbCO_{3(s)} \quad (6.3)$$

The enthalpy change for this reaction is practically impossible to measure directly but a Hess's law calculation allows the value for ΔH_1 to be calculated from standard enthalpy of formation values.

Q3.

a) Write an equation for the standard enthalpy of formation of lead II oxide, including state symbols, and use it to give a definition for **the standard enthalpy of formation.**
b) Use the values for the standard enthalpies of formation in Table 6.1 to calculate a value for ΔH_1 for eqn 6.3. Use should use an enthalpy diagram to illustrate your answer.

Table 6.1 Standard enthalpies of formation for PbO, CO_2 and $PbCO_3$.

Compound	ΔH^{θ}_{f}/kJ mol^{-1}
PbO	−217.3
CO_2	−393.5
$PbCO_3$	−700.0

6.4 THE ROMANS AND LEAD

It's a bit of a hoary old chestnut with historians that the downfall of the Roman Empire was all down to lead: lead pipes, lead acetate as a sweetener in wine, lead cosmetics, *etc.*, but there may well be *some* truth in it. There is no doubt that the Romans smelted lead on such a large scale that the evidence can be seen in ice cores from the Greenland ice caps. They were also well aware that the slaves who were sent down the mines to extract the lead had brief and unpleasant lives, but you didn't get to run the greatest empire the world had yet seen by putting slave welfare high on the agenda.

Lead is too soft for weapons but it can easily be beaten into sheets and folded around wooden rods to make lead pipes. Using lead mined from Spain and Britain, the Romans set about building the most extensive public water supply system of their era. But the local water quality proved to be an important factor.

Q4.

a) A sample of "soft" tap water was found to contain 0.3 ppm of Pb^{2+}. Express this concentration in mol dm^{-3}.
b) A sample of hard water was found to be saturated with $CaCO_3$, $K_{sp} = 6.0 \times 10^{-9}$ mol^2 dm^{-6}. Calculate the concentration of carbonate ions in this solution in mol dm^{-3}.
c) Given the K_{sp} for $PbCO_3 = 7.4 \times 10^{-14}$ mol $^2dm^{-6}$, calculate the maximum quantity in ppm of Pb^{2+} that would be soluble in this sample of hard water.
d) Given your answers to parts a–c, comment on the effect of water hardness in areas with lead pipes.

6.4.1 Roman Wine

With the lack of quality control that we take for granted today, Roman wine must have been of very variable vintage and the oxidation of ethanol to ethanoic acid would have made it taste, quite literally, like vinegar. Either by design or accident, it was discovered that the acid taste of the wine could be compensated for by adding "sugar of lead" or lead II ethanoate. Initially, it may well have been recognized that wine from lead cups tasted sweeter. After that, sugar of lead was made by design to sweeten all manner of foods. This would have been one of the occasions in history when the *Hoi Poloi* were largely spared, as excessive consumption of wine and finer foods would certainly have only been within the reach of the "great and the good". The steady decline of the Julio–Claudian dynasty, starting with the impotent Julius and ending with the matricidal Nero, seems to give the lead theory a whiff of plausibility.

6.5 LEAD IN YOUR BONES?

If you were to measure the concentration of lead in various body tissues, you would find that it is typically present at 0.2 ppm in blood, 0.2–3.0 ppm in soft body tissue and 3–30 ppm in bone. Why the difference? *In vivo*, biochemistry has a problem distinguishing Ca^{2+} and Pb^{2+} ions (more on this later). They both form +2 cations and the ionic radius of lead is only about 20% larger than that of calcium. Consequently lead has a habit of substituting itself for calcium and, thus, it is not surprising to find it turning up in bones. Some studies have even shown that calcium supplements derived from organic sources, such as oyster shells, may in fact contain high levels of lead.

6.6 RED LEAD

Red lead (an oxide of lead) can be used as a pigment and paint primer. Its formula is generally given as Pb_3O_4. However, the reaction of this compound with nitric acid reveals a little more (eqn 6.4):

$$Pb_3O_{4(s)} + 4HNO_{3(aq)} \rightarrow PbO_{2(s)} + 2Pb(NO_3)_{2(aq)} + 2H_2O_{(l)} \quad (6.4)$$

Q5.

a) Calculate the mass of lead IV oxide that you would expect to collect if excess nitric acid was added to 17.13 g of Pb_3O_4 and you recovered 88% of the solid.
b) Would you describe this as a redox reaction? Explain your answer.

6.7 THE LEAD-ACID ACCUMULATOR BATTERY

In many commercial areas the usage of lead and its compounds is in decline and it is now banned in most paints. However, the lead-acid accumulator battery is still an essential part of most cars and, in this industry, demand is on the increase with approximately half of all lead extracted used in this way. Lead can form compounds with oxidation states +2 and +4 and it is this ability that is utilized in the car battery. The two relevant half-cell equations are shown is eqn 6.5 and 6.6.

$$Pb^{2+}{}_{(aq)} + 2e^- \rightarrow Pb_{(s)} \qquad E^\theta = -0.13 \text{ eV} \tag{6.5}$$

$$PbO_{2(s)} + 4H^+{}_{(aq)} + 2e^- \rightarrow Pb^{2+}{}_{(aq)} + 4H_2O_{(l)} \qquad E^\theta = +1.47 \text{ eV} \tag{6.6}$$

Q6.

a) What is meant by a secondary cell in this context?
b) Use the two half-cell equations (eqn 6.5 and 6.6) to construct an equation representing the overall reaction for the cell discharge and calculate the E^θ cell.

6.8 LEAD ISOTOPES AND THE AGE OF THE EARTH

We have seen that isotopes have proven to be a very useful "tool" in many areas of science, with half-lives ranging from fractions of a second to about a third of the age of the universe. The relative abundance of lead isotopes 206 and 207 has helped in giving an approximate age for the Earth. The two most common isotopes of uranium (^{238}U and ^{235}U) are both radioactive and decay with half-lives of 4.5×10^9 years and 7.0×10^8 years, respectively. Uranium is incorporated into the mineral zircon ($ZrSiO_4$), which, rather fortuitously, excludes lead atoms from its crystalline structure when it is formed. The end product of the decay series for ^{238}U is the lead isotope ^{206}Pb, whereas the end product in the decay series for ^{235}U is ^{207}Pb. The $^{238}U/^{206}Pb$ and the $^{235}U/^{207}Pb$ ratios are used as independent "clocks" for the dating of zircon minerals. Both decay plots give an approximate Earth age of 4.55×10^9 years $\pm$ 1%.

Q7. A third *radioactive* isotope of Lead, ^{210}Pb, decays by β emission. Write an equation for this reaction and identify the element produced.

6.9 TETRA-ETHYL LEAD

With the benefit of hindsight, the use of tetra-ethyl lead in petrol looks like one of the more reckless decisions in recent history. There is no doubt that, as an additive to petrol, it improved octane rating (fuel performance) and reduced wear and tear on valves in the engine, but as early as the 1920s there were clear indications of its negative impact on health.

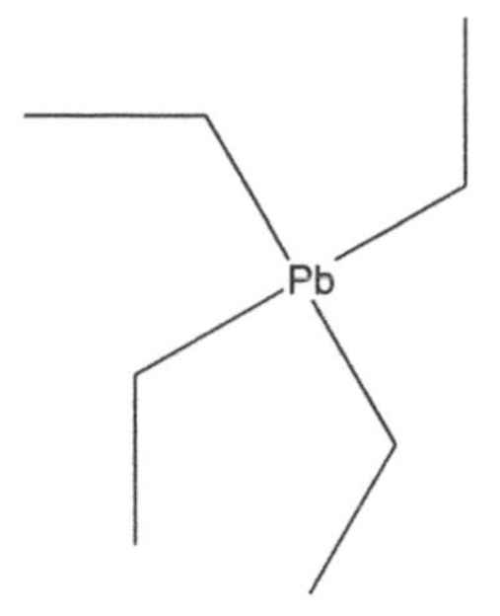

Figure 6.2 Structure of tetra-ethyl lead.

Q8. The skeletal structure of tetra-ethyl lead is shown in Figure 6.2.

a) Give the molecular formula of the molecule.
b) Suggest a value for the C–Pb–C bond angle, giving a reason for your choice.

Q9. A human blood sample showed a lead ion concentration of 0.166 mg dm^{-3}. Covert this value to:

a) Concentration in μmol dm^{-3}.
b) Concentration in ppm.

Although a blood level sample of 0.166 mg dm^{-3} may be typical, it hides the fact that samples taken from an urban population may be ten times greater than a rural population and, even more notable, this value may be up to 500 times that of pre-industrial levels. Is this important? Well some studies are now beginning to report some startling results. In a paper by Rick Nevin[1] the following conclusion was drawn:

"The findings with respect to violent crime are also consistent with studies indicating that children with higher bone lead tend to display more aggressive and delinquent behavior. This analysis demonstrates that widespread exposure to lead is likely to have profound implications for a wide array of socially undesirable outcomes."[1]

Another conclusion from a second paper is equally striking:

"Prenatal and postnatal blood lead concentrations are associated with higher rates of total arrests and/or arrests for offenses involving violence. This is the first prospective study to demonstrate an

association between developmental exposure to lead and adult criminal behavior."[2]

Depending upon your viewpoint this may indicate either a) the immense political influence that the petrochemical industry had or still has in the West or b) our collective willingness to ignore stuff we'd rather not see.

6.10 LEAD POISONING: SOME MORE BIOCHEMISTRY

Recent advances in molecular imaging using nuclear magnetic resonance (NMR) and X-ray crystallography have enabled scientists to find out exactly how lead can disrupt things at the molecular level. We have already noted that lead can displace calcium in many important binding proteins and, given calcium's importance in cellular signalling, it is not surprising that chronic lead exposure can have such a negative impact. It is also now known that lead will displace zinc ions from an important set of molecular transcription factors. In these proteins the lead ion will form dative bonds between two histidine and two cysteine amino acids. The resultant conformational changes in these proteins can have a disastrous impact on one of the most important cellular functions: DNA transcription!

Q10. The two amino acids, cysteine and histidine, are shown in Figure 6.3 (left and right, respectively).

a) Identify the chiral carbon in each molecule.
b) Which of the two molecules is more likely to have a significant ultraviolet absorbance at 280 nm? Explain your choice.
c) Suggest which parts of these two molecules are responsible for forming a stable co-ordination complex with a lead ion.

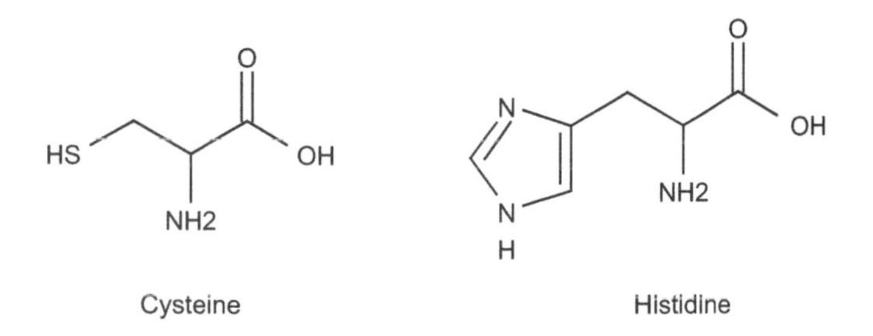

Figure 6.3 Structures of cysteine (left) and histidine (right).

6.11 CONCLUSIONS

From the earliest observations in Roman times to contemporary statistical analyses, lead has been associated with detrimental effects on health and there is an increasing amount of evidence to show clear causal relationships. Faced with this evidence, governments in many parts of the world have responded by first reducing and then banning lead additives in petrol. However, in one respect lead is rather important: it is an excellent shield for gamma rays. Should we ever have to deal with radioactive fall-out, then you can expect lead to be in demand!

REFERENCES

1. R. Nevin, *et al.*, Lead Poisoning and Juvenile Crime Update, 2013. Available online at: http://www.ricknevin.com/uploads/Lead_Poisoning_and_Juvenile_Crime_Update.pdf. Accessed 5/12/2013.
2. J. P. Wright, K. N. Dietrich, M. D. Ris, R. W. Hornung and S. D. Wessel *et al.*, Association of Prenatal and Childhood Blood Lead Concentrations with Criminal Arrests in Early Adulthood, *PLoS Med.*, 2008, 5(5), e101.

Answers to questions in this chapter can be found on pages 184–186.

CHAPTER 7

Lithium

Chances are that you are not a million miles away from a laptop, tablet or a smart phone as you read this—maybe you're even reading it on one. Be that the case, then many of the lithium atoms in the battery powering these devises will be some of the oldest atoms in the universe. Most lithium atoms, together with hydrogen and helium atoms, were created at the Big Bang event, currently estimated to be about 13.7 billion years ago. Some of those atoms eventually condensed, with many others formed by nucleo-synthesis reactions in stars to form our planet. Today, one of the greatest demands for the element is for its use in electrochemical cells and it is lithium's low density and highly negative standard electrode potential that makes it so useful for this purpose. In addition, it is an important constituent of lightweight metal alloys and its compounds are used in both chemical synthesis and the pharmaceutical industry. This chapter will also give us the chance to look at some important unifying concepts in physical chemistry.

7.1 LITHIUM AND MAGNESIUM: A DIAGONAL RELATIONSHIP

Diagonal relationships are seldom mentioned in A level syllabuses these days but the idea seems logical enough and possibly even quite helpful. Metal reactivity increases going down groups I and II, but decreases going across the period. Thus, going across from Li to Be is

Around the World in 18 Elements
By David A. Scott

Published by the Royal Society of Chemistry, www.rsc.org

decreasing the reactivity, but going down from Be to Mg increases it again. The net effect of the two changes results in a similar reactivity.

Q1.

a) Give the electronic s,p,d notation for atoms of the elements Li and Mg.
b) Write equations for i) the first ionization energy of Li and ii) the second ionization energy of Mg.
c) The metallic radii of Li and Mg are 0.157 and 0.160 nm, respectively, even though they are in different periods of the periodic table. Suggest why this is so.
d) Suggest two other elements in the periodic table likely to show a diagonal relationship.

7.2 STONES AND STARS

Lithium is named after the Greek word *lithos*, meaning stone, and the lithosphere is the name given to the ridged outer crust of the Earth. You might, therefore, expect lithium to be one of the more abundant elements in the Earth's crust. At 20 ppm, however, it is less abundant than the rare earth metal cerium, which as you can see is not particularly rare. The element was identified in 1817 by the Swedish chemist J. A. Arfvedson (1792–1841), but it wasn't until 1821 that a sample of the element was isolated. This might beg the question: how could it be identified *before* it was isolated? To answer this question we need to appreciate the significance of the atomic emission and absorbance spectra, which have revolutionized our understanding of the universe.

In earlier times the noble metals were often adulterated with baser metals for obvious economic reasons. If nitric acid was added to gold and it fizzed, well then you could be sure that you didn't have pure gold (it was almost certainly adulterated with copper)! To be certain that you actually had gold, you needed to handle a sample to test, weigh, feel, prod, add acid to it and so on. In 1835 the French philosopher August Comte stated that humans would never be able to understand the chemical composition of stars. The stars were simply too far away and it was inconceivable that samples would ever become available for analysis in the sense analogous to the testing of gold. However, two years after Comte's death Kirchhoff and Bunsen (yes that one) developed a high-temperature gas burner in which they tested various substances, observing that many gave flames of characteristic colours. Kirchhoff and Bunsen went on to realize that the characteristic flame colours were indicative of particular elements and, when analyzed by a spectroscope, each element had a characteristic line spectrum that was unique to that element.

It was Arfvedson that first noticed a characteristic crimson flame in 1817 when analyzing a sample of the ore petalite ($LiAlSi_4O_{10}$) and suggested the existence of a hitherto unknown element that was a lighter version of the element sodium. It was to be a further 37 years before a sufficiently large sample was isolated by electrolysis to enable its physical properties to be measured.

Today, the elemental composition of stars is determined from atomic emission and absorbance spectra and, indeed, stars are now even classified on the basis of their elemental composition. In an ironic twist, however, some astronomical observations suggest that there seems to be rather less lithium in the universe than the Big Bang model predicts. Will the missing lithium be found or will the current theoretical model need to be reviewed? Only time will tell. Such is the constant discourse between scientific theory and observation that makes science mankind's pre-eminent achievement.

Q2. One of the most intense lines in the atomic emission spectrum of lithium is at 670.8 nm. Calculate the energy in J of a photon of energy of this wavelength.

Q3.

a) The relative atomic mass of a sample of lithium is 6.94. Assuming that this relative atomic mass is due to the disproportionate abundance of two lithium isotopes, ^6Li and ^7Li, calculate the percentage of each isotope in the sample.
b) Would you expect the lithium isotopes to have different atomic emission line spectra? Explain your answer.

7.3 THE ELECTROCHEMISTRY OF LITHIUM

It is lithium's electrochemical properties that will probably see demand for the metal continue to increase for the foreseeable future. With a density of 0.53 g cm^{-3}, lithium's advantage over denser metals, such as nickel and cadmium, is clear. However, it is lithium's standard electrode potential that is key to its use in electrochemical cells and it is worth taking some time to consider what appears to be something of anomalous value (Table 7.1).

Table 7.1 Standard electrode potentials for Li, Na, K and Rb.

Metal	*Standard electrode potential/eV*
Li	−3.03
Na	−2.71
K	−2.92
Rb	−2.93

Table 7.2 Definitions of the variables in eqn 7.1.

Symbol	*Definition*	*What it means*
ΔG	Gibb's free energy change	A negative value means the reaction is feasible
ΔH	Enthalpy change	The heat energy change associated with a reaction
T	Absolute temperature	The temperature in Kelvin (*i.e.*, °C + 273)
ΔS	Entropy change of the system	A measure of disorder in the system under consideration
$-n$	Number of electrons transferred	
F	The Faraday constant	Taken as 96 500 C. This is the total electric charge carried by 1 mole of electrons
E	The electrode potential	Measured relative to the hydrogen electrode
R	Universal gas constant	Boltzmann's constant × Avogadro's constant
$\ln K$	Natural logarithm of the equilibrium constant	

As a student of chemistry many years ago, I remember being rather troubled by the anomalous value for the electrode potential of lithium. The increasingly negative value for the electrode potential from Na to Rb made perfect sense: the increasing reactivity, observed in the reaction of the metals with water fitted, with the explanation that the more reactive metal has a greater tendency to release electrons, making the metal a stronger reducing agent. But lithium clearly doesn't fit. Being told that the first member of a group often exhibited anomalous properties didn't answer the question, it merely restated it.

In the following section I am going to use the relationships shown in eqn 7.1. The symbols are defined in Table 7.2.

$$\Delta G = \Delta H - T\Delta S = -nFe = -RT\ln K \tag{7.1}$$

Eqn 7.1 links a great deal of physical chemistry and I encourage my students to learn it by heart. I hope to show that, by inspecting this equation and the numbers generated, we can gain an insight as to what is happening at the atomic level. To start, let's focus on the relationship between the enthalpy of reaction and the electrode potential (eqn 7.2).

$$-nFE = \Delta H - T\Delta S \tag{7.2}$$

This immediately reminds us that the electrode potential is proportional not just to the enthalpy change but to the enthalpy *and* the entropy change of the system. Given that, let's now input some figures for comparing sodium and lithium.

For lithium:

$$E = 3.03 \text{ eV and } \Delta H_f \text{ Li}^+{}_{(aq)} = -278.5 \text{ kJ mol}^{-1}.$$

From these data we can calculate the entropy change of the system when 1 mole of lithium metal forms 1 mole of aqueous lithium ions:

$$-1 \times 96\,500 \times 3.03 = -278\,500 - (298 \times \Delta S)$$

$$-292\,395 = -278\,500 - (298 \times \Delta S)$$

$$-13\,895 = -298 \times \Delta S$$

$$\Delta S = +46.6 \text{ J K}^{-1}.$$

Q4. Calculate the ΔS_{system} value for sodium using the following data:

$$E = 2.71 \text{ eV and } \Delta H_f \text{ Na}^+{}_{(aq)} = -240.1 \text{ kJ mol}^{-1}.$$

Hopefully, you will have calculated a larger value compared to the ΔS_{system} for lithium, showing that the process results in a greater increase in disorder.

From the calculation it is clear that the change in entropy for the formation of the hydrated lithium ion is small in comparison to sodium and the other alkali metals. Having shown this quantitatively we might now ask why? The answer is due to the high charge density of the Li^+ ion. This will result in a high enthalpy of hydration for the ion and probably form not just one hydration shell but a second surrounding the first. The increase in order that this generates means that the overall entropy change is smaller than might be expected. This, then, is part of the reason why the electrode potential of lithium is so strongly reducing and, combined with its low density, makes it so attractive for electrochemical cells.

7.4 LITHIUM CARBONATE

Like calcium carbonate, lithium carbonate shows some interesting trends in solubility with temperature change (Table 7.3).

Q5. Plot a graph of the temperature of solution *vs.* the solubility of solution and use it to predict a solubility value at 40 °C.

Can we explain why the solubility of Li_2CO_3 *decreases* with temperature? Again, a consideration of free energy and entropy is revealing. The relationship between free energy and the solubility product of lithium carbonate is shown below:

$$K_{sp} = [Li^+]^2[CO_3^{2-}] = 8.5\times10^{-4}\ mol^3\ dm^{-9}.$$

$$\Delta G^0 = -RT\ln K_{sp}.$$

Inputting values for $T = 298$ K, we can calculate the free energy change for the solubility of lithium carbonate.

$$\Delta G^0 = -8.314\times298\times(\ln 8.5\times10^{-4}) = 17517\ J\ or\ 17.5\ kJ\ mol^{-1}.$$

The positive value tells us that the forward reaction is less feasible than the reverse.

What happens as temperature rises? Let's input values for $T = 333$ K (60 °C):

$$\Delta G^0 = -8.314\times333\times(\ln 8.5\times10^{-4}) = 19.6\ kJ\ mol^{-1}.$$

The more positive value at 333 K tells us that the reaction becomes less feasible at higher temperatures, which is what we observe from our data.

Substituting ΔG^0 for $\Delta H^0 - T\Delta S^0$, we get: $\Delta H^0 - T\Delta S^0 = -RT\ln K_{sp}$.

Table 7.3 The solubility of lithium carbonate with temperature.

Temperature of solution/°C	*Solubility of solution/g per 100 cm³*
0	1.54
20	1.32
60	1.01
100	0.72

$\Delta H^0 - T\Delta S^0$ is positive at 298 K. If it becomes *more* positive as T increases this can only mean that the entropy change for Li_2CO_3 dissolving in water is **negative**! In other words: an increase in order. How can this be explained?

When considering the entropy change as a substance dissolves, we must remember to consider the whole system *including the water*. The solid lithium carbonate has a low entropy but water is a disordered liquid and has a relatively high entropy. The key here is that, when lithium carbonate dissolves, the loss of order from the solid is more than compensated for by the increase in order of the water molecules hydrating the lithium and carbonate ions. Again, the high charge density of the Li^+ ion means that there is probably not one but two shells of hydration around the ion, resulting in a decrease in entropy for those water molecules.

So, we begin to see that many of lithium's anomalous properties seem to be related to its relatively high charge density. This high ionic charge density has implications for the thermal stability of lithium nitrate and lithium carbonate.

Q6.

a) Write balanced equations for the thermal decomposition of lithium nitrate and lithium carbonate.
b) Suggest why the behavior of these compounds is more like the equivalent magnesium compound in group II than that for sodium in group I.

7.5 LITHIUM COMPOUNDS IN ORGANIC SYNTHESIS

One of the most important reagents in synthetic organic chemistry is lithium aluminium hydride, $LiAlH_4$. It can be synthesized by reacting lithium hydride with aluminium chloride, giving lithium chloride as a by product.

Q7.

a) Write a balanced equation for the reaction of lithium hydride with aluminium chloride to produce the products stated above.
b) What is the atom economy for this reaction?
c) What mass of aluminium chloride is required to react with 10 kg of LiH? Give your answer to the nearest whole kg.

d) Draw the structure of the AlH_4^- ion, showing all valence electrons and bond angles.

Lithium aluminium hydride is one of the most potent reducing agents but it has the disadvantage of reacting violently with water. As a result, a non aqueous solvent, such as ethoxyethane, is generally used.

Q8.

a) $LiAlH_4$ reacts with water to produce aluminium hydroxide, lithium hydroxide and hydrogen. Write a balanced equation for this reaction.
b) What mass of $LiAlH_4$ would be needed to generate one m^3 of hydrogen gas?
c) What implications does this reaction have for the storage of $LiAlH_4$?

Q9.

a) Complete the table below for the products of the reaction between the stated organic reactant and $LiAlH_4$.

Organic reactant formula	*Name of reactant*	*Product formula after reacting with $LiAlH_4$ in ethoxyethane*	*Name of product*
$CH_3CH_2CH_2CH_2CH_2Cl$			
$CH_3CH_2CH_2OH$			
$(C_2H_5)CHOH$			
$CH_3CH_2CONH_2$			
CH_3CH_2COOH			
CH_3COCH_2COOH			

b) The reduction of the final product in the table above produces a racemic mixture. Draw out the displayed formula of the reduced product and identify the chiral carbon.
c) How might you use infrared (IR) spectroscopy to follow the progress of the reaction in b)? Explain your answer.

7.6 APOLLO 13: LITHIUM HYDROXIDE TO THE RESCUE

The rescue of astronauts Lovell, Haise and Swiggert from the Apollo 13 accident was referred to as "NASAs finest hour" by flight director

Gene Kranz, and few would argue that the collective ingenuity, skill and courage shown by all involved was remarkable. One of the many problems that the crew had to deal with on their return was the removal of the excess carbon dioxide in the confined space of the lunar module. This involved making connections between the carbon dioxide scrubbers, which, for design reasons, were incompatible between the lunar and command modules. From a chemical point of view, a question might be: why use lithium hydroxide instead of the more common sodium hydroxide? The answer has to do with the cost of launching a payload into space. Prices range from $3000–$30 000 per kg.

Q10.

a) Write balanced chemical equations for the reactions of sodium hydroxide and lithium hydroxide with carbon dioxide.
b) Calculate the mass of each hydroxide required to fully absorb 50 dm^3 of carbon dioxide. (Assume 1 mole of gas occupies 24 dm^3.)
c) How much carbon dioxide can be absorbed per kg of lithium hydroxide and sodium hydroxide?

7.7 LITHIUM: "LIKE WALKING ON THE BOTTOM OF THE OCEAN"

Nobody is quite sure how lithium works as a treatment for bipolar disorder but there seems little doubt that it does. In 1949 John Cade, a senior medical officer in Victoria, Australia, who was experimenting on guinea pigs, found that these normally excitable animals were rendered rather docile on having a solution of a lithium salt injected into their blood stream. Since that time, lithium salts have been used to control the extreme mood swings associated with bipolar disorder. However, many report a strange sensation of being detached from the world and consider it a price too great to pay. In addition, the levels of lithium in the blood have to be carefully regulated as the difference between therapeutic and dangerous levels can be small.

Q11.

a) When lithium carbonate enters the stomach, it reacts with the hydrochloric acid to produce lithium chloride. Write a balanced equation for this reaction.

b) Blood serum concentrations of the lithium ion should not exceed 2.5 mmol dm^{-3}. Calculate the maximum mass of lithium dissolved in 100 cm^3 of blood serum.

7.8 CONCLUSIONS

In Aldous Huxley's dystopian novel, *Brave New World*, people moderate their moods with Soma. Some research has even shown a correlation between lower levels of suicide and naturally raised levels of lithium in the water supply. Who knows, one day lithium fluoride may be added to the water supply to improve moods *and* strengthen teeth (see Chapter 10).

Soma: "All the advantages of Christianity and alcohol; non of their defects."[1]

Henry Foster in Aldous Huxley's *Brave New World*

REFERENCE

1. A. Huxley, *Brave New World*, 1932, Chatto and Windus, London.

Answers to questions in this chapter can be found on pages 187–190.

CHAPTER 8

Iodine

Iodine can be found in group VII and period five in the periodic table. With a relative atomic mass of about 127, it is perhaps the most massive atom that is essential to complex animal life. It may be that organisms had been flourishing for some time before iodine was incorporated into their biochemistry. One immediate question, then, is why do we need it at all—especially when prokaryotes and plants seem to do just fine without it? This, of course, is one of the mysteries of evolution. There is no design behind the use of iodine and its compounds. At some point, an organism will have stumbled upon it and found a use for it. Conferring some advantage on the organism, iodine chemistry will have been retained and modified over the eons. Today, iodine chemistry has an impact far greater than its earthly abundance might suggest.

In addition to other functions, iodine is widely used in quantitative analysis (and it is a particular favorite for A level examiners)! There are some questions in this section that will familiarize you with calculations of this type.

8.1 A NEW ELEMENT FROM SEAWEED

Not for the first time we find that international conflict is the moving force behind progress in chemistry. Bernard Courtois (1777–1838) is acknowledged as the discoverer of iodine. Whilst trying to find a

Around the World in 18 Elements
By David A. Scott

Published by the Royal Society of Chemistry, www.rsc.org

source of nitrates for Napoleon's war in Europe, Courtois accidentally spilled concentrated sulfuric acid on the ash recovered from seaweed and is said to have seen a purple vapour, which we would now recognize as iodine (eqn 8.1).

$$8I^- + 10H^+ + SO_4^{2-} \rightarrow H_2S + 4I_2 + 4H_2O \tag{8.1}$$

Whether Courtois also noted the unmistakably foul smell of hydrogen sulfide is not recorded.

Q1. Identify the oxidizing and reducing agent in this reaction (eqn 8.1) and give oxidation numbers for all species involved.

8.2 TESTING FOR IODINE WITH STARCH

We all know the test for iodine: add starch solution and a blue/black colour appears, but the chemistry behind this reaction remained rather unclear for some time. Firstly, you don't get the blue/black colour with starch if the iodine is dissolved in a non-polar solvent, such as cyclohexane! Today, the general consensus for the observed reaction seems to be as follows:

1. Molecular iodine reacts with iodide to form tri-iodide (eqn 8.2).

$$I_{2(aq)} + I^-_{(aq)} \rightleftharpoons I_3^-{}_{(aq)} \tag{8.2}$$

2. Further reactions can occur to form the penta-iodide ion (eqn 8.3) and so on to form poly-ions with 7, 9, 11, *etc.* iodine atoms.

$$I_{2(aq)} + I_3^-{}_{(aq)} \rightleftharpoons I_5^-{}_{(aq)} \tag{8.3}$$

3. These poly-iodide ions then interact with amylose units (which are short, coiled polymers of glucose monomers present within starch).
4. The complex in which the poly-iodide fits inside the amylose coil absorbs nearly all visible light, and thus appears blue/black.
5. However, the complex is fairly temperature sensitive and heating the solution will decompose the complex and the colour will disappear.

Q2. In eqn 8.1 the iodide ion acts as a Lewis base. State what is meant by a Lewis base and describe the arrangement of the electrons around the central iodine in I_3^-.

8.3 EXTRACTING IODINE FROM SEAWEED

At the beginning of the 2012 school year, a very keen student of mine approached me with the idea of investigating an element as part of a silver CREST award (run by the British Science Association, BSA). The idea was to research and present a biography of the element with a little bit of practical work thrown in. After some discussion, we agreed on iodine, aware of the fact that it can be extracted from certain types of seaweed. There was no shortage of protocols available on the internet, so we set about it, collecting a species of *Laminaria* (kelp) from the East Kent coast. The first thing that we both learned is that you need to get a decent heat-resistant crucible for the all-important "ashing" process; this process having dispatched various bits of inferior lab kit! It is worthwhile learning early on that chemistry really is a practical subject and no matter how straight forward a laboratory procedure is there **is no substitute for actually doing it!**

I say this here for the following reason: in some quarters there is a belief that the virtual world of the internet will offer suitable and adequate simulations of chemical processes to the extent that practical investigations may somehow become, at best, an optional extra or, at worst, obsolete. This is not the case!

Success eventually arrived after we had purchased a special variety of kelp from Ireland and we were able to convert the iodide in the seaweed ash into iodine by reaction with hydrogen peroxide. We subsequently quantified the amount of iodine using a thiosulfate titration with a starch indicator. You can process our results for yourself in Q3 and Q4.

Q3. The equation for acidified hydrogen peroxide acting as an oxidizing agent is shown in eqn 8.4:

$$2H^{+} + H_2O_2 + 2e^{-} \rightarrow 2H_2O \quad (8.4)$$

Write a half equation for the oxidation of iodide ions to iodine and give the full balanced equation for the redox reaction.

Q4. A 10.0 g dried sample of *Laminaria* was burned to ash and then reacted with excess acidified hydrogen peroxide. The excess hydrogen peroxide was then decomposed by a catalytic reaction with manganese IV oxide. The catalyst was then removed by filtration. The resulting solution was made up to a volume of 500 cm^3. A 25 cm^3 sample of this solution required 10.05 cm^3 of 0.001 M $Na_2S_2O_{3(aq)}$ to reach an end point with starch indicator (eqn 8.5).

$$I_2 + 2S_2O_3{}^{2-} \rightarrow 2I^{-} + S_4O_6{}^{2-} \quad (8.5)$$

Use this information to calculate the percentage mass of iodine in this sample of *Laminaria*.

8.4 IODINE DEFICIENCY

It is estimated that up to two billion people worldwide suffer a degree of iodine deficiency.[1] This deficiency is thought to give rise to "the most preventable form of mental impairment worldwide."[1] The importance of iodine in the diet and the major attempts to address this dire situation are considered in the following sections.

8.4.1 Thyroxine

The bulk of iodine in the human body can be found covalently bonded to carbon in the hormone thyroxine, which consists of a mixture of thyroid hormones, T3 and T4 (T4 is shown in Figure 8.1). T3 differs from T4 in having one less iodine atom on the phenol group.

Q5.

a) Calculate the percentage mass of iodine in T4.
b) Explain the significance of the H atom shown bonded with a dashed wedge in Figure 8.1. What implications might this have for the synthetic production of T4?
c) Suggest and draw possible molecular structures for T4 in aqueous solutions of i) pH 3 and ii) pH 11.

Thyroxine plays a key role in the body, regulating metabolic rate and effecting rates of protein synthesis. A lack of iodine in the diet will result in a reduction of thyroxine production, which can have a number of physiological effects, including a swelling of the thyroid tissue called a goitre. In an attempt to treat chronic iodine deficiency, iodine is added to salt in many parts of the world either in the form of sodium iodide or sodium/potassium iodate V (see Q6).

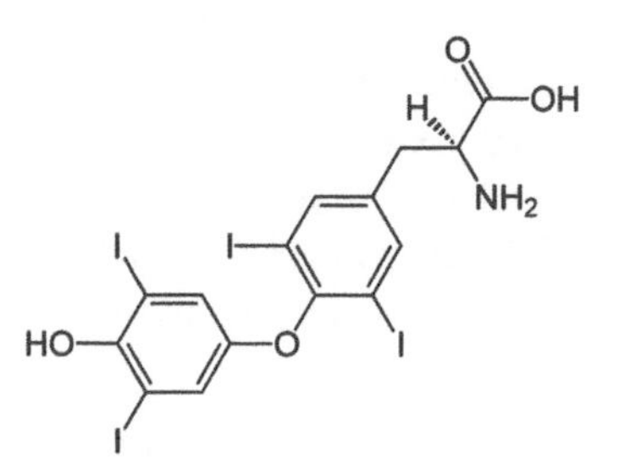

Figure 8.1 Structure of T4: a thyroid hormone.

Q6. As part of the investigation into iodine, we decided to quantify the amount of sodium iodate V ($NaIO_3$) present in a brand of iodized salt. The packet instructions specified that the salt contained "no less than 1150 μg of iodine per 100 g of salt". We dissolved 50 g of salt in distilled water, added excess acidified $KI_{(aq)}$ in order to reduce the iodate V ions to molecular iodine and made the solution up to 250 cm^3 (eqn 8.6).

$$IO_3^- + 6I^- + 6H^+ \rightarrow 3I_2 + 3H_2O \quad (8.6)$$

We then titrated 50 cm^3 of our reduced solution with 0.002 M $Na_2S_2O_3$ solution and recorded an average end point of 4.85 cm^3. Calculate the actual amount of iodine in 100 g of salt. How does it compare to the stated quantity on the packet?

Prior to the use of iodate V salts, iodine was added to salt in the form of sodium iodide (NaI) but it became clear that, over time, the iodide ions reacted with oxygen and carbon dioxide in the air according to eqn 8.7:

$$4NaI + O_2 + 2CO_2 \rightarrow 2Na_2CO_3 + 2I_2 \quad (8.7)$$

As a result, sodium or potassium iodate V is now the preferred iodine additive to salt.

Q7.

a) Identify the species that is oxidized and the species that is reduced in eqn 8.7.
b) Suggest a chemical test and the expected result that you might use to show that the above reaction had taken place.
c) Suggest why iodate V ions are more stable in air.

8.5 IODINE AND RADIOACTIVITY

In 1986 a nuclear reactor situated north of Kiev in what was then the USSR exploded, releasing a number of radioactive isotopes, one of which was iodine-131. This had serious implications for people in the fall-out area. Firstly iodine-131 has a half life of just over eight days, and secondly the isotope decays by β and γ emission. As we have already seen, the body absorbs iodine easily and the bulk of it is incorporated into thyroid hormones. An immediate solution was to offer exposed individuals sodium iodide pills. By saturating the body with the non-radioactive isotope, the body would simply excrete the excess iodine, including any radioactive isotopes. This happened in Poland (but not sadly in the Ukraine or Belarus). Nearly 30 years after the event, epidemiological studies indicate a significant increased incidence of thyroid cancers in individuals who were teenagers at the time and resident in the fall-out zone.

Radioactive isotopes can also be used to help *treat* certain types of thyroid cancer. Iodine-131 is concentrated in thyroid tissue and so the radiation is concentrated in the area of need. The penetration of the β radiation is less than 10 mm into the thyroid tissue and, thus, the effect is localized.

Q8. A dosage of 3.7×10^{12} Bq (Bq = Becquerel) is given orally to a patient. This value represents the initial activity of the ^{131}I isotope. Given that ^{131}I has a half-life of 8.04 days, use eqn 8.8 to calculate the activity that remains after four weeks.

$$A = A_0\, e^{-(0.693/T_{1/2})} \tag{8.8}$$

8.6 IODIDE AS A CATALYST

With the exception of fluorine, all group VII elements (the halogens) can show variable oxidation states, and elemental iodine can react to form compounds showing six different oxidation states. The ability to alternate between oxidation states often allows an element to show catalytic properties (see the chapters on copper (Chapter 9), manganese (Chapter 16) and iron (Chapter 2)) and the decomposition of hydrogen peroxide is catalyzed by iodide ions as shown in eqn 8.9:

$$2H_2O_2 \xrightarrow{I^-} 2H_2O + O_2 \tag{8.9}$$

Q9.

a) An investigation into the kinetics of this reaction (eqn 8.9) suggests that it is **first order** with respect to hydrogen peroxide **and** iodide ions. Write down an expression for the rate equation.
b) Use the equation to calculate units for the rate constant.
c) Further studies suggest that IO^- is an intermediate species in a two-step reaction mechanism. Give a two-step reaction mechanism that is consistent with the information given and suggest which step is the rate determining step, giving a reason for your answer.

8.7 IODINE, KELP AND THE CLIMATE

Given that animals need a source of iodine in their diet, it might be surprising that it is a *plant* that is far and away the strongest accumulator of iodine in the living world. As we have already seen, it is the *Laminariales* (or, more commonly, kelp—a species of brown seaweed) that are the great iodine scavengers. It has only recently become clear that the presence of iodine in kelp (mostly in the form of iodide) has both a protective function for the plant and may also play a significant role in atmospheric chemistry and the climate.

Recent research has shown that, when subjected to oxidative stress, iodide ions are released into the intracellular matrix of kelp and help act as reducing agents. (We have already seen how iodide can catalyze the disproportionation of hydrogen peroxide to water and oxygen.) Oxidizing agents can be the cause of damage to DNA and other macromolecules. In addition to this function, the presence of iodine monoxide (IO) in the atmosphere above kelp beds is thought to act as a precursor for cloud condensation nuclei. Climactic modeling is a notoriously complex problem and the role of the iodine biogeochemical cycle has only recently been factored in. As with all systems that are fundamentally chaotic, it is very difficult, perhaps impossible, to accurately quantify their impact.

Q10. Draw a dot and cross diagram of iodine monoxide and use it to explain why you would expect it to be kinetically unstable.

8.8 CONCLUSIONS

As the 64th most abundant element in the Earth's crust, we might be forgiven for assuming that iodine's impact would be minimal, but I hope that this chapter has suggested it is anything but. It is something of a mystery as to why such a rare element has ended up with such a central role in a hormone as important as thyroxine. Williams and Frausto da Silva have stated that iodinated phenols are used as protective poisons by many organisms.[2] Once drawn into an organism's biochemistry, evolution has a habit of using extant molecules in novel and unexpected ways.

REFERENCES

1. M. B. Zimmermann, P. L. Jooste and C. S. Pandav, *The Lancet*, 2008, **372**(9645), 1251–1262.
2. R. J. P. Williams and J. J. R. Frausto da Silva, *The Natural Selection of the Chemical Elements*, 1996, Clarendon Press, Oxford.

Answers to the questions in this chapter can be found on pages 191–192.

CHAPTER 9

Copper

It may not be immediately obvious why a chapter about copper should begin with a picture of a horseshoe crab (Figure 9.1). The reason lies with this primitive animal's blood, which is **blue** when oxygenated. The blue colour is due to the presence of two copper atoms in the protein haemocyanin, which are oxidized from Cu^{+} to Cu^{2+} when oxygen molecules bond. As a complex metalloprotein, it is structurally very different from haemoglobin and illustrates how evolution can come up with very different solutions to the same problem; in this case, transporting oxygen for aerobic respiration.

Q1.

a) Draw "electrons in boxes" diagrams for Cu^{+} and Cu^{2+} ions.
b) Explain why the oxidation of Cu^{+} to Cu^{2+} in haemocyanin results in a change from colourless to blue.

Copper makes its first appearance in human culture at about 5000 BCE and was the first metal to be smelted. It was the major partner in defining the Bronze Age and, today, with the globally increasing desire for electrical technology, demand for copper has never been greater. In addition to looking at the historical development of copper usage, this chapter will enable us to consider some of the typical properties exhibited by transition metals and their compounds. From this, we will move to think about how these typical transition metal properties are exploited and managed by biological systems.

Around the World in 18 Elements
By David A. Scott

Published by the Royal Society of Chemistry, www.rsc.org

Figure 9.1 A horseshoe crab.[1] Reproduced with permission from David Hodgeson ©.

9.1 COPPER IN HISTORY

The Great Pyramid of Giza, made for the Pharaoh Cheops, would be a remarkable achievement even if it was built today. It is all the more remarkable when you consider that the limestone blocks were probably shaped by copper tools. Copper is a relatively soft metal and would have required frequent sharpening. Tin, with which copper is alloyed to make the harder bronze, had yet to be discovered. At this stage in history, copper was still rare and may have been known only as the native metal.

The Minoan civilization flourished between 1600–1100 BCE on the island of Crete and was a sophisticated bronze-age culture. There is evidence that the culture sourced much of its copper from the island of Cyprus. The Latin for Cyprus is *Cuprum* and it is from this name that we get the symbol for copper, Cu. The presence of Minoan pottery in Cyprus hints at what once would have been strong trade links between the two islands.

Q2. Explain why alloying copper with tin results in a harder substance.

The discovery of how to extract copper metal from its ore may well have happened by accident. The ore malachite, $Cu_2CO_3(OH)_2$, has a striking green/blue colour and may well have been prized for jewellery. It is possible that a piece of malachite may have fallen into a wood and charcoal fire and gradually been reduced to copper metal. A large enough fire with sufficient heat and carbon may have enabled the following reactions to take place (eqn 9.1 and 9.2):

$$Cu_2CO_3(OH)_2 \rightarrow 2CuO + H_2O + CO_2 \quad (9.1)$$

$$CuO + C \rightarrow 2Cu + CO_2 \quad (9.2)$$

Q3.

a) Looking at eqn 9.1 and 9.2, identify which is the thermal decomposition reaction and which is the redox reaction.
b) What mass of copper could be extracted from 500 kg of pure malachite?
c) What is the atom economy of the second step in the reaction?

The British scientist and science fiction writer Arthur C. Clarke once said:

"Any sufficiently advanced technology is indistinguishable from magic."

For the early metal workers, extracting the reddish metal from a blue/green stone must really have seemed magical. We can, perhaps, see in this reaction the beginnings of alchemy and for many hundreds of years to come this kind of "magical" transmutation of matter would be controlled by a quasi priestly cast of metallurgists. Ritual would have been an important part of the smelting process. Without written instructions or any kind of time piece, the fidelity of the process would have been ensured by the close following of a ritual or the singing of a song, much as the Japanese tea ceremony ensures a good cup of tea.

9.2 VENUS, GODDESS OF LOVE (AND COPPER)

We saw in the chapter on lead (Chapter 6) that there were seven metals known to the ancient civilizations, each associated with a planet. Copper was associated with Venus perhaps, in part, because its slight greenish tinge in the night sky was reminiscent of the green patina that copper gets as it gradually oxidizes and then reacts with carbon dioxide in the air (Figure 9.2).

Figure 9.2 Alchemical symbol for copper.

For the alchemists, copper represented a transitionary state between the base metal lead and the most noble gold. Metals were thought to be evolving in the ground and the alchemists' attempts to convert base metals into gold were simply conceived as a way of speeding up a natural process. There is evidence that, in medieval times, mines were closed for a period of time to allow them to "regenerate" new metals. This may seem silly by today's scientific standards but the idea of leaving a field fallow for a couple of years was well established and allowing the mine to recuperate† was a logical extension of the same idea.

The Romans had another technique for getting copper. They would put scrap iron in rivers and streams near mines and leave it for months at a time. If the river water contained a sufficient concentration of copper II ions then it was simply a matter of waiting until the copper metal built up on the outside of the iron.

Q4. Use the electrode potentials in eqn 9.3 and 9.4 to help explain the chemistry behind the extraction of copper metal this way.

$$Fe^{2+}{}_{(aq)} + 2e^{-} \rightarrow Fe_{(s)} \quad E^{\theta} = -0.44 \text{ eV} \tag{9.3}$$

$$Cu^{2+}{}_{(aq)} + 2e^{-} \rightarrow Cu_{(s)} \quad E^{\theta} = +0.34 \text{ eV} \tag{9.4}$$

9.3 MODERN COPPER REFINING

Although copper has many uses, it is its properties as an electrical conductor for which it is most prized. In the vast majority of cases, metals are alloyed to improve properties but copper used in electrical wiring must be 99.995% pure as the resistivity of copper drops off quickly as the percentage impurity increases.

If we were to consider the composition of the Earth's crust (including plants, animals, cities, *etc.*), we would find that over 90% is made from just seventeen elements: O, Si, Al, Fe, Ca, Na, K, Mg, H, Ti, Cl, P, Mn, C, S, N and F. You will immediately see that copper is not one of the seventeen and, if it were evenly distributed throughout the Earth's crust, then at 50 ppm it would not be economically viable to extract. However, the dynamic nature of our Earth's crust, driven by convection currents in the magma beneath, causes minerals and ores to be concentrated and associated with certain geological features.

Given the energy and time required to extract the copper, miners would not currently consider an ore much less than 20 000 ppm (but,

†As keen as I was to find an etymological link between *recuperate* and *cuprum* sadly I could find none.

of course, market fluctuations and demand will be a factor here). For purposes of illustration, let's consider the ore **bornite** (Cu_5FeS_4):

$$2Cu_5FeS_4 + 4O_2 \rightarrow 5Cu_2S + 2FeO + 3SO_2 \quad (9.5)$$

$$Cu_2S + O_2 \rightarrow 2Cu + SO_2 \quad (9.6)$$

Q5.

a) Calculate the percentage mass of copper in the ore bornite.
b) Calculate the total volume of SO_2 gas (measured at room temperature/pressure) produced per 1000 kg of copper produced. Take the molar gas volume to be 24 dm^3 at room temperature/pressure and give your answer to the nearest m^3.

At this stage, the copper is 99% pure, which is insufficient for electrical purposes. The final stage involves electrolysis. The impure copper is made into an anode and immersed into a solution of acidified copper II sulphate. Cathodes are made from stainless steel. As the reaction proceeds copper metal at the anode is oxidized and dissolves in the electrolyte. At the cathode, copper II ions are reduced back to copper. One of the impurities of copper at this stage is silver, which forms a precipitate with the sulfate ions in the electrolyte and is deposited at the bottom of the electrolytic cell.

c) If a current of 200 A was passed through an industrial electrolytic cell for 18 h, what mass of copper would you expect to collect at the cathode? (1 Faraday = 96 500 C).
d) Suggest why the actual amount of copper collected is likely to be less than your answer in c).

9.4 COPPER COINS

One perennial favourite demonstration to younger pupils studying chemistry is the two pence piece in concentrated nitric acid (see Chapter 3 for the equations). The reaction is extremely exothermic, gives off thick brown fumes and leaves a dark blue solution, and so must be carried out in a fume cupboard. What remains at the end of the reaction depends upon the age of your two pence piece as, since 1992, two pence coins have changed from being solid bronze to copper-plated steel. The older coin can be completely oxidized, leaving no solid provided you have enough nitric acid. The post 1992 coins will leave a steel core at the end of the reaction. The change was necessitated by the increase in copper prices on the world market.

At one point, market prices meant that there was nearly three pence worth of copper in a two pence coin!

9.5 COPPER, REDOX REACTIONS AND THE CONCEPT OF FREE ENERGY

Redox reactions tend toward a state of equilibrium and thus, in moving from a state of dis-equilibrium to equilibrium, can make free energy available to do work. There is a simple relationship between the enthalpy change for a reaction and the free energy change, and thus an estimation of the enthalpy change of a reaction is the first step to determining the free energy change of a reaction.

9.5.1 Experimental Procedure for Determining Enthalpy of Reaction

An experimental procedure for measuring enthalpy of reaction is given below. Read through the protocol and answer the questions that follow. (Although a sketch graph is provided, it is worthwhile plotting the results as accurately as you can and extrapolating lines of best fit yourself.)

1. Pipette 50.0 cm^3 of 1.00 M $CuSO_{4(aq)}$ into a beaker.
2. Measure the initial temperature of the copper II sulphate solution to the nearest 0.1 °C and continue recording the temperature every 30 s for 2 min.
3. Weigh out between 3.5–4.0 g zinc powder.
4. At 2.5 min, add all of the zinc powder to the copper sulphate solution and stir with the thermometer.
5. Record the temperature every 30 s for the next 3 min.

9.5.2 Results

Time/s	*Temperature/°C*
0	21.2
30	21.3
60	21.2
90	21.2
120	21.2
150 (zinc added)	21.2
180	60.5
210	60.0
240	59.5
270	59.1
300	58.6
330	58.1

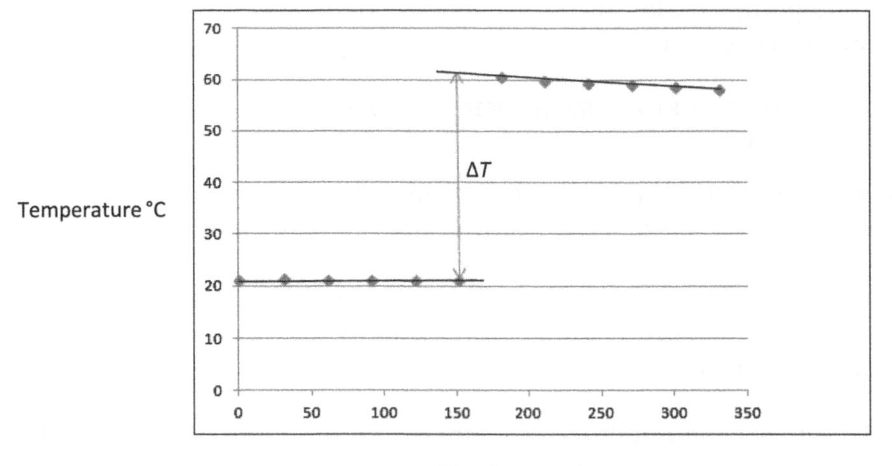

Figure 9.3 A plot of time (s) against temperature (°C) to determine the enthalpy of reaction.

You should plot your data and extrapolate the lines of best fit so as to give an estimated value for the maximum temperature rise. The illustration in Figure 9.3 shows such a plot.

Extrapolated maximum temperature = 61.0 °C, giving a temperature change of 61.0 – 21.2 = 39.8 °C = ΔT.

Q6.

a) Write an ionic equation for the reaction taking place between zinc and the copper II sulfate solution.
b) Show that the amount of zinc powder used in the protocol is in excess.
c) Using the value of ΔT (above), calculate a value for the enthalpy of reaction. You may assume that the specific heat capacity of 1.00 M copper II sulfate is 4.2 kJ kg^{-1} °C^{-1}.
d) Compare your calculated value with a data book value of ΔH = -217 kJ mol^{-1} and suggest reasons for any discrepancy.

The value for the enthalpy change calculated is related to free energy change (ΔG, also known as Gibb's free energy change) by the following relationship (eqn 9.7):

$$\Delta G = \Delta H - T\Delta S \tag{9.7}$$

Q7. Look back at the reaction you wrote for Q6.a) and suggest why the values of ΔG and ΔH are likely to be fairly close in this particular reaction.

The relationship between free energy and electrode potential is given by eqn 9.8:

$$\Delta G = -nFE, \tag{9.8}$$

where n is the number of electrons transferred and F is Faraday's constant (96 500 C).

Q8.

a) Use the following standard electrode potentials (eqn 9.9 and 9.10) to calculate the electrode potential generated by combining the zinc and copper half cells.

$$Cu^{2+}{}_{(aq)} + 2e^{-} \rightleftharpoons Cu(s) \quad E^{\theta} = +0.34\ eV \tag{9.9}$$

$$Zn^{2+}{}_{(aq)} + 2e^{-} \rightleftharpoons Zn(s) \quad E^{\theta} = -0.76\ eV \tag{9.10}$$

b) Hence, show that the free energy value calculated using $\Delta G = -nFE$ is consistent with the experimental value calculated above and the data book value.

9.6 COPPER AND BIOCHEMISTRY

In Robert A. Heinlein's sci-fi classic *Stranger in a Strange Land*[3] the author refers to life as *"that oddity of distorted entropy."* Although this might be a level of reductionism too extreme for all but the most determined materialist, there is no doubt that life does involve coupling chemical reactions in such a way as to prevent its systems reaching complete equilibrium or, to put it more bluntly, death. So it might not come as a total surprise to see copper and zinc coupled in the enzyme *superoxide dismutase.*

The job of this enzyme is to deal with the very reactive superoxide species O_2^{-}, a product of oxidative metabolism, which, if left unchecked, could do considerable intracellular damage. The overall catalyzed reaction can be represented by eqn 9.11:

$$2O_2^{-} + 2H^{+} \rightarrow O_2 + H_2O_2 \tag{9.11}$$

According to Creighton,[2] the copper and zinc ions are co-ordinated to a histidine residue at the catalytic site of the enzyme (Figure 9.4).

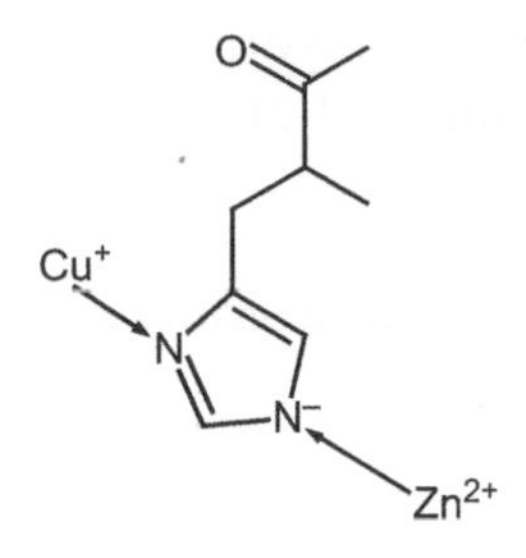

Figure 9.4 Copper and zinc ions co-ordinated to histidine.

The key step in the reaction with the superoxide anion involves the ability of copper II to accept an electron, generating an O_2 molecule, followed by a release of the electron with $2H^+$ and a second O_2^- to generate H_2O_2.

Q9.

a) Assign oxidation numbers to all oxygen species in eqn 9.11.
b) What name is given to this type of redox reaction?
c) What typical transition metal property is copper exhibiting in this catalytic reaction?

9.6.1 Copper as a Micronutrient

A typical human adult may have a total of about 75 mg of copper in all of their body tissues and, although this may not seem like much, the inability to either retain or excrete copper adequately has major consequences. Copper retention manifests itself as Wilson's disease, which can result in problems of the brain and the liver. Menkes disease results from copper deficiency and, like Wilson's disease, it is a recessive trait; however, unlike Wilson's disease, it is linked to the X chromosome and consequently much more common in males. This is because, having only one X chromosome, males carrying the mutation will certainly have the disease. Females, with two X chromosomes, would have to have both X chromosomes with the mutation in order to have the disease, which is statistically far less probable. In this disease copper is deficient in liver and brain tissue but may accumulate in the kidneys. Both diseases can be fatal if not treated adequately and highlight just how important copper is for normal bodily function.

Q10. One clinical test for Wilson's disease is to measure the amount of copper passed in 24 h. More than 100 mg passed in 24 h is a clear indicator of Wilson's disease. A patient passed 1850 cm^3 in a 24 h period, which contained 78 mg copper. Calculate this concentration in μmol dm^{-3}.

9.7 CONCLUSIONS

Once upon a time, wooden ship hulls would be sheathed with copper to prevent decay by marine organisms. As such, these ships fared rather better than those that were unprotected, the literal meaning of which developed into an expression conveying a "sure thing" (a high degree of security), *i.e.*, a "copper-bottomed investment". Could the next copper-bottomed investment be...investing in copper? Some might think so! In February 2011 the value of copper topped $10 000 per tonne and, with developing economies continuing to grow, demand is unlikely to drop for long. So, if you've got a few coppers to spare...

REFERENCES

1. Image from http://theliquidearth.org/wp-content/uploads/2009/10/horseshoe_crab.jpg.
2. T. E. Creighton, *Proteins: Structures and Molecular Properties*, 1997, 2nd edn, Freeman, London, p. 362.
3. R. A. Heinlein, *Stranger in a Strange Land*, 1961, Putnam publishing group, New York.

Answers to questions in this chapter can be found on pages 193–195.

CHAPTER 10

Fluorine

Fluorine is, in every sense, a difficult and potentially confusing element, not least because many students still spell it flourine! Here are just some of the problems likely to be encountered in a classroom:

Teacher: "Fluorine is an extremely reactive non-metal."
Pupil: "But isn't it in toothpaste?"
Teacher:"Yes it is. . .I'll explain later."

And then further on up the school. . .

Pupil: "You mean to say that when fluorine reacts with oxygen it's the **oxygen** that's oxidized and not the fluorine..?!"

Or. . .

Pupil: "If hydrofluoric acid reacts with glass. . .it must be a really strong acid!"
Teacher: "Err. . .well no, it's weaker than phosphoric acid, the stuff you get in a cola drink actually."

. . .. you get the idea.

In order to resolve these apparent contradictions we need to take a good look at fluorine. Practically, however, this is most unlikely to happen in the laboratory as fluorine is, as already stated, extremely reactive. This does not mean that fluorine is without its uses; far from

Around the World in 18 Elements
By David A. Scott

Published by the Royal Society of Chemistry, www.rsc.org

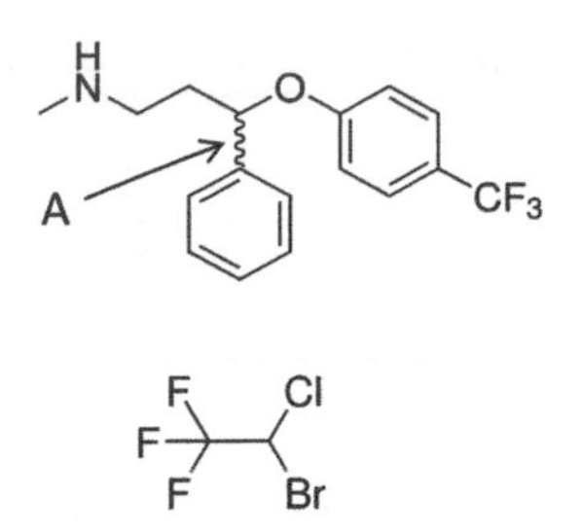

Figure 10.1 Compound 1 (top) and compound 2 (bottom).

it! It is because the carbon–fluorine bond is almost unknown in nature and, consequently, a whole section of organofluorine chemistry has developed, which has had a major impact on the pharmaceutical, agrochemical and polymer industries. There are lots of opportunities in this chapter to carry out some web searches on fluorine-based molecules.

Two synthetic organofluorine compounds are shown in Figure 10.1.

Q1.

a) Carry out a web search and identify each of the compounds in Figure 10.1 by their common name. What are they used for? (It is helpful to have a molecular drawing package that enables you to draw and name the molecules.)
b) Give the IUPAC name for compound 2.
c) Work out the molecular mass and percentage mass of F in each compound.
d) Explain the use of the wavy line, labeled A, in compound 1.
e) Fluorine has only one naturally occurring isotope: ^{19}F. However, chlorine has two stable isotopes: ^{35}Cl and ^{37}Cl, which occur with abundances of 75% and 25%, respectively, and bromine has two isotopes: ^{79}Br and ^{81}Br, with approximately 50% abundance each. Use this information to explain why compound 2 gives *three* molecular ion peaks at *m*/*z* values of 196, 198 and 200 in the approximate ratio of 3 : 4 : 1.
f) Why might the ratio of these peaks differ slightly in a real sample?

10.1 NATURAL ABUNDANCE

Fluorine is surprisingly abundant in the Earth's crust. At number thirteen in the abundance list, it is more common than sulfur and

Table 10.1 Solubility of CaF_2 and $CaCl_2$ in water.

Compound	*Solubility in water/mol per 100 g*
CaF_2	2.31×10^{-5}
$CaCl_2$	5.36×10^{-1}

chlorine; however, due to its reactivity, it is never found as the free element.† Unlike sulfur, which is concentrated by geological activity, fluorine is largely dispersed in a range of rock types. However, the relative presence of chloride and fluoride ions in sea water is striking: there are typically 14 000 chloride ions for every fluoride ion in sea water, which poses the interesting geological question: where is all the fluorine? The obvious answer seems to be that it is locked up in insoluble compounds in hard-wearing rocks, such as basalts and granites.

Consider the data in Table 10.1 and then answer the questions that follow.

Q2.

a) Calculate the **mass** of each compound that could dissolve in 100 g of pure water.
b) How might you expect the presence of calcium chloride in solution to affect the subsequent solubility of calcium fluoride? Explain your answer.

10.2 APATITES

Apatites are minerals with the general formula $Ca_5(PO_4)_3X$, where X can be OH, Cl or F. Fluorapatite (X = F) is the most resistant of the three mineral acids. The primary chemical component of teeth enamel is hydroxyapatite (X = OH) and the practice of water fluoridation has been carried out in various parts of the world for some time. The argument in favour is based on the fact that a gradual replacement of hydroxide with fluoride will result in stronger teeth, which are more resistant to tooth decay. The most cursory scan of the internet will show water fluoridation to be a real bone of contention, with water

†Most fluoride ions are indeed locked up in insoluble compounds but some recent research suggests that diatomic fluorine may exist trapped in calcium fluoride minerals generated by the effect of β radiation from the decay of unstable isotopes present in calcium fluoride.

fluoridation supposedly being the cause of things as diverse as colon cancer and brain damage.

Q3. Carry out a web search on the pros and cons of water fluoridation. How do you go about judging the validity and reliability of a web resource?

In the UK the department for rural affairs permits a maximum of 1.5 mg L^{-1} fluoride in drinking water and leaves the fluoridation decision to local health authorities.

10.3 DETERMINING THE BOND ENTHALPY OF F_2 USING A BORN–HABER CYCLE

Use the data in Table 10.2 to answer the questions below.

Q4.

a) Construct a Born–Haber cycle for calcium fluoride and use it to calculate an enthalpy value for the F–F bond.
b) Compare the bond enthalpy you have calculated to the other diatomic halogens and comment and suggest an explanation for any trend.

10.4 FLUORINE AND REDOX

Early definitions of redox reactions are based on the idea of oxygen (or "the acidifying principle of the air", as Lavoisier called it) combining with some substances and being removed from others. Combustion would be an example of oxidation, whereas the smelting of iron from its ore is an example of reduction. As the development of the atomic

Table 10.2 Enthalpy changes for the formation of calcium fluoride.

Reaction	*Enthalpy change/kJ mol^{-1}*
Atomisation of Ca	178.2
First ionization energy of Ca	590.0
Second ionization energy of Ca	1145.0
First electron affinity of F	−328.0
Lattice enthalpy of CaF_2	−2630.0
Enthalpy of formation of CaF_2	−1219.6

model progressed, and protons, electrons and neutrons were discovered, a broader definition of redox in terms of electrons was developed to encompass a wider range of reactions. Thus, removal of electrons became the definition of oxidation and the addition of electrons became the definition of reduction. Nevertheless, the terms *oxidation* or *reduction* were retained and from this idea came the idea of oxidation number.

Oxidation numbers are easily assigned for ionic compounds (the numbers are the same as the ionic charges) but in covalent compounds the concept of electronegativity becomes important. The element with the greater electronegativity value is conceived to have the greater share of the bonding electrons and, consequently, has the negative electronegative value.

Q5.

a) Give the IUPAC name and assign oxidation numbers to the elements underlined in each of the following compounds:

i. $Ca\underline{S}O_3$
ii. $K_2\underline{Mn}O_4$
iii. $(NH_4)_2\underline{Mo}O_4$
iv. $\underline{Cl}_2O_7$
v. $\underline{P}_4O_{10}$
vi. $\underline{C}Cl_4$
vii. $\underline{F}_2O$

Hopefully, you will have recognized that oxygen has a negative oxidation number **except** when it is bonded with fluorine, where it has an oxidation number of +2. This is because fluorine is the only element that is more electronegative than oxygen.

b) Assign oxidation numbers to all the reactants and products in the reaction shown in eqn 10.1:

$$2F_2 + 2H_2O \rightarrow 4HF + O_2 \quad (10.1)$$

10.5 ORGANOFLUORINE COMPOUNDS

It turns out that the high electronegativity value for fluorine is one of the properties that make it so useful in pharmaceutical molecules. It goes without saying that a pharmaceutical molecule must tackle the symptoms for which it is designed, but that is far from the only problem that must be considered. One of the body's immediate responses to a foreign molecule is to attempt to dispose of it by breaking it down. A molecule could be the most effective painkiller ever discovered but if it lasted no more than a few minutes in the blood stream it would have no chance to get to work. Fluorine can deactivate a molecule by drawing electrons away from a potential reaction site, making it less susceptible to electrophiles.

Q6.

a) Use the internet to find the structural formula for each of the following molecules:

i) Atorvastatin
ii) Voriconazole
iii) Paroxetine

b) What do you notice about the position of the fluorine atom(s) in each of the molecules in part a)?

10.6 NORFLOXACIN: AN ANTIBIOTIC OF LAST RESORT?

The mention of super-bugs can be something of a tabloid favourite on a slow news day. What will happen when the antibiotic stalwarts no longer work? At present, a group of molecules called fluoroquinolones—of which norfloxacin (Figure 10.2) is one—are held back as antibiotics of last resort. They are thought to interfere with bacterial enzymes called DNA gyrases, which are essential in the reproduction of bacterial DNA. Enzymes called topoisomerases are used for the analogous process in eukaryotes and are unaffected by fluoroquinolones.

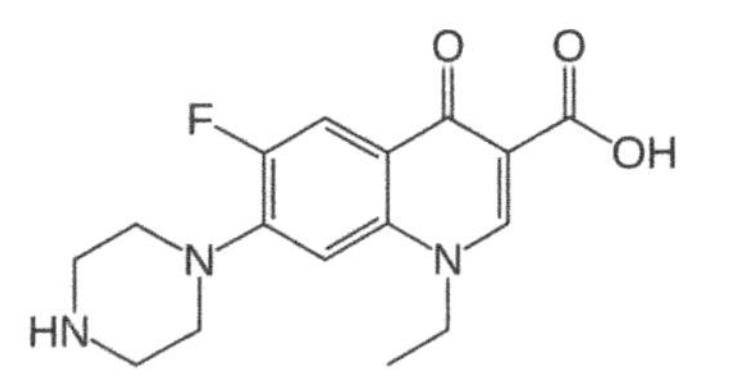

Figure 10.2 Structure of norfloxacin.

Q7.

a) Identify the following functional groups in norfloxacin:

i) Carboxylic acid
ii) Ketone
iii) Secondary amine
iv) Tertiary amine

b) How many H atoms are there in a molecule of norfloxacin and how many **different** proton environments would you expect to see in a ^{1}H NMR spectrum?

10.7 FLUORO POLYMERS

The C–F bond is the strongest covalent bond in carbon chemistry, and thus fluorocarbons are very unreactive. They are also hydrophobic. Both factors have implications for their uses. The fluorine atom has a covalent radius less than twice that of hydrogen but, unlike hydrogen, it has three closely held lone pairs of electrons in its valence shell. The strongly hydrophobic nature of organofluorines is exhibited by polytetrafluoroethene (PTFE), marketed as Teflon on non-stick frying pans. This polymer is analogous to polythene in which all hydrogen atoms are replaced by fluorine atoms. Indeed, the main problem of using PTFE was overcoming the difficulty of getting it to stick to the frying pan itself! Another plus is the high heat stability of PTFE and it is even used as a coating on high velocity bullets.

10.7.1 Inertness can be a Problem

Naturally occurring polymers have the advantage of being biodegradable. Cellulose, for example, is one of Earth's oldest and most common polymers. It is not surprising that bacteria have evolved to be able to break down cellulose and metabolize the glucose sugars released. Man-made polymers are a different story. PTFE was first made in 1938 and, as a result, there are no naturally occurring microbes that are able to break down the polymer. This has one profound consequence; namely, such polymers tend to accumulate in the environment.

Perfluoro-octanoic acid (PFOA) is another fluorinated compound that accumulates in the environment and, again, is extremely resistant to further metabolism. It is the fluorinated analogue of octanoic acid in which all of the alkyl hydrogens are replaced by fluorine atoms.

Q8.

a) Draw the skeletal structure of PFOA.
b) The Environmental Protection Agency (EPA) has defined a maximum limit of 400 ng dm^{-3} PFOA in drinking water. If a person consumes 3.5 dm^3 of drinking water in a day, what is the theoretical maximum amount of PFOA consumed in mol?

10.8 THE EFFECT OF FLUORINE ON pK_a

The replacement of one methyl hydrogen with a fluorine atom in a molecule of ethanoic acid may seem to be a relatively insignificant change to the molecule but it is the electronegativity difference between hydrogen and fluorine that makes all the difference. Consider the pK_a values of the two acids, shown in Table 10.3.

Q9.

a) Explain why the presence of a fluorine atom has such a significant impact on the pK_a value of fluoroethanoic acid.
b) Calculate the pH of a 0.01 M solution of ethanoic acid stating any assumptions made.
c) Calculate the pH of a 0.01 M solution of fluoroethanoic acid. (Note that you will need to solve a quadratic equation in this case.)
d) Explain why you cannot make the same assumptions in part c) as you can in part b).

10.9 SULFUR HEXAFLUORIDE

Sulfur hexafluoride is a very stable molecule with octahedral geometry and is widely used where a dense unreactive gas is needed, such as

Table 10.3 pK_a values in ethanoic acid and fluoroethanoic acid.

Acid	*pK_a value*
Ethanoic	4.76
Fluoroethanoic	2.66

high voltage electrical circuitry and the electrolytic extraction of magnesium from molten magnesium chloride. Coincidentally, about 2500 years ago, Plato made the association between the octahedron (one of the five Platonic solids) and the "air" element. Not for the first time in this book the association between stability and symmetry is noted!

Q10.

a) Draw a molecule of SF_6 showing all bond angles.
b) Assign an oxidation number to sulfur in SF_6 and explain the term "expanded octet" in relation to this molecule.

10.10 SARIN

Sarin is one of the more notorious fluorine-containing compounds due, in part, to its use by the Japanese religious sect Aum Shinrikyo on the Tokyo underground in 1994. This highly toxic nerve agent resulted in the death of thirteen people. The structure is shownin Figure 10.3.

Q11.

a) Assign an oxidation number to phosphorus in sarin and draw electrons in boxes to show how this number is achieved.
b) Identify the chiral centre of the structure of sarin shown in Figure 10.3.

10.11 ^{19}F ATOMS CAN GIVE RISE TO AN NMR SPECTRUM

Most A level syllabuses cover the basics of NMR spectroscopy, focusing on the ^{1}H nucleus or, in some cases, the ^{13}C nucleus. In addition to having some important chemical implications for synthetic molecules, the ^{19}F nucleus also gives a strong NMR resonance, which is very sensitive to its local environment. Amino acids labeled with ^{19}F atoms can give structural information (Figure 10.4).

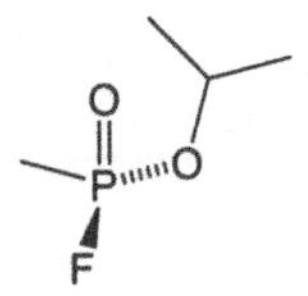

Figure 10.3 Structure of sarin.

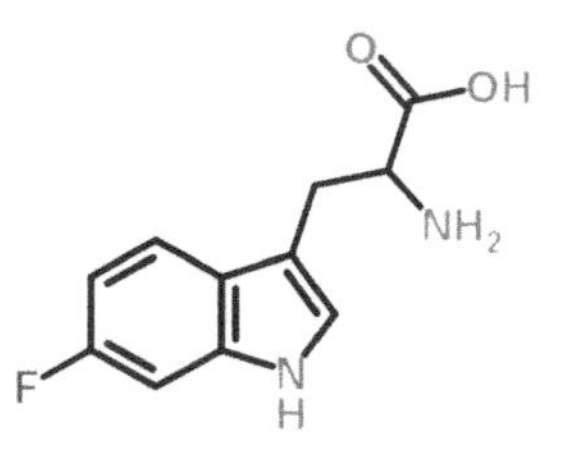

Figure 10.4 Structure of 6-fluorotryptophan.

Q12. The ^{19}F spin is sensitive to its immediate environment. Can you suggest why ^{19}F resonances may be able to give information about the tertiary state of a protein?

10.12 SOME MICROBES CAN METABOLIZE ORGANOFLUORINES

It is never a good idea to make sweeping generalizations when it comes to biochemistry. The C–F bond may be *almost* unknown in the natural world, as indicated at the outset of this chapter, but the exceptions are becoming an area of great interest. A South African plant *Dichaptetalum cymosum* has been linked with a number of livestock fatalities due to the production of fluoroethanoic acid and, more recently, a strain of *Streptomyces* soil bacterium has been shown to contain genes that code for a fluorinase enzyme that can combine inorganic fluoride to make a fluorinated ribose sugar. At present, many fluorinated organic derivatives are made using gaseous fluorine as a reagent with all the associated health and safety issues. A set of enzymes that enable the biocatalytic synthesis of organic derivatives that use fluoride as the starting material would be very desirable indeed. In addition, the problem of non-biodegradability could be solved by the controlled use of bacterial organisms that havc bccn engineered to break down organofluorines.

10.13 CONCLUSIONS

Fluorine chemistry is an area of research that has burgeoned over the last 50 years or so. There are now whole journals focusing on fluorine chemistry and a great number of pharmaceuticals and agrochemical now have fluorine somewhere in the mix. Nature may have barely touched the potential of fluorine chemistry but humans have chosen to do so and have achieved some notable successes. We may need many more!

Answers to question in this chapter are on pages 196–200.

CHAPTER 11

Aluminium

We are all familiar with the Bronze and Iron ages as periods in history when these particular metals led to advances in technology that defined the dominant cultures. It's possible that, sometime in the future, historians may look back on a period between 1830 and 1970 as the Aluminium Age. I admit to choosing fairly arbitrary dates but then many would argue that the Bronze and Iron ages are equally arbitrarily defined. What is certainly true is that the technology that enabled the large scale extraction of aluminium from its ore had a huge impact on technology and, consequently, society. The UK's electricity supply network (the National Grid) would be inconceivable without aluminium, as would international air transport to the extent that it now makes a journey to the other side of the world in less than 24 hours possible. I was recently reminded that it was only 66 years between the Wright brothers' first flight and Armstrong and Aldrin setting foot on the moon; both impossible without aluminium.

Aluminium is the third metal in period three of the periodic table and has both low density and good electrical conductivity. Its compounds have been used as mordants from early times and, today, aluminum compounds are important reagents in organic synthesis. Aluminium is also the third most abundant element in the Earth's crust but it has no known biological role. Evolution seems to have rejected aluminium. Indeed, it has been implicated in some diseases, such as Alzheimer's.

Around the World in 18 Elements
By David A. Scott

Published by the Royal Society of Chemistry, www.rsc.org

11.1 ALUMINIUM: PERIOD THREE'S ODD MAN OUT

Going across period 3 in the periodic table you come, in order, to sodium then magnesium, both of which are essential to life; there is the sodium/potassium pump in cell membranes and magnesium adopts a central position in chlorophyll. Then comes aluminium, which has no known biological role. Silicon is important in silicate exoskeletons found in many aquatic organisms, as well as grasses, and the remaining non-metals (phosphorus, sulfur and chlorine) have central roles in all life-forms in one or more forms. "Lazy" argon, like all noble gases, has no chemistry to speak of. So why, of all the period three elements, is aluminium so neglected? As the third most abundant element in the Earth's crust, it is not for lack of availability. For some reason, the evolution of life left it out of the mix until the 1800s, when the discovery of electrolysis enabled its extraction. Genes may have ignored aluminium but memes[†] have made extensive use of it.

11.2 ALUMINIUM EXTRACTION

By now we should be familiar with the idea that a metal can be abundant in the Earth's crust as a compound but unavailable as the free metal due to its chemical reactivity. Aluminium makes up, on average, 8.1% of the crust and is commercially extracted from bauxite, an ore named after a place in the French district of Les Baux-de-Provence near Arles from where it was first mined.

Bauxite consists of hydrated aluminum oxide together with impurities, including silicates, iron oxides and titanium oxide. The first procedure in the extraction of aluminium from bauxite is to remove the impurities. To do this we make use of aluminium's amphoteric nature.

In aqueous solution aluminium oxide is insoluble but in a moderately concentrated sodium hydroxide solution the following reaction takes place (eqn 11.1):

$$Al_2O_3 \cdot 3H_2O_{(s)} + 2NaOH_{(aq)} \rightarrow 2NaAl(OH)_{4(aq)} \qquad (11.1)$$

The sodium aluminate species is soluble in a sodium hydroxide solution of this concentration but iron III oxide and silicon dioxide are not. Consequently, the mixture can now be filtered and the insoluble impurities removed. The sodium aluminate solution is then

[†]A meme is an aspect of culture that is passed from one individual to another by non-genetics means.

treated with acid and the hydrated aluminium oxide precipitates out, is filtered and dried. Further heating drives off water to leave pure aluminium oxide.

Q1.

a) Explain what is meant by the term *amphoteric*.
b) Write balanced equations for the reactions between i) aluminium oxide and hydrochloric acid and ii) aluminium oxide and sodium hydroxide.
c) Give the electronic configuration of Al and Al^{3+}. With which element is Al^{3+} **isoelectronic**? Explain your answer.
d) Draw the shape of the $[Al(OH)_4]^-$ ion and explain its shape in terms of orbital hybridization.
e) What volume of 6.0 M NaOH would be needed to completely dissolve 50 kg of hydrated aluminium oxide?

11.2.1 Electrolysis

The electrolytic extraction of aluminium from bauxite is familiar to all GCSE chemistry students. In electrolytic cells aluminium ions are reduced to aluminium and oxide ions are oxidized to oxygen. Initially, the technical problem was how to do the process economically. Aluminium oxide's melting point of 2072 °C meant that the energy required to melt the oxide, and thus mobilize the ions, was considerable and that meant its price per kg was very high. When the King of Siam visited Louis-Napoléon Bonaparte III in France in the 19th century, he wasn't palmed off with any old gold or platinum trinket but instead given a watch charm of aluminium. Napoléon himself was very keen to find an economic way of extracting aluminum as he felt that it would give his troops the edge in the Franco–Prussian war. Sadly, it didn't, and Otto von Bismarck went on to unify Germany. The rest, as they say, is history and a fair bit of geography and politics to boot!

The solution (glossing over the pun) came with a rare mineral called cryolite (Na_3AlF_6), which was first discovered in Greenland. It was Charles Hall in the USA and Paul Heroult in France who, in 1886, separately realized that aluminium oxide could dissolve in cryolite, a compound that melts at the relatively modest temperature of 950–1000 °C. This made the process far more economically viable and eventually led to the large-scale extraction of aluminium. In a sad coincidence both men were to die at the relatively early age of 51 in 1914. It is more than probable that exposure to fluorine gas contributed to the demise of both. Although oxygen is the main product at the anode during the electrolytic process, side reactions can give rise to other products, such as fluorine gas. In the days before health and safety both men would almost certainly have been exposed to excessive amounts.

From 1939 to 1943, during World War II, aluminium production rocketed to over two million tonnes but, afterwards, production dropped and there was something of an oversupply of aluminium. Consequently, there were attempts to build prefabricated houses from aluminium for a while but, unsurprisingly, these proved rather unpopular, not least an electrical hazard.

Q2.

a) An aluminium electrolytic cell has a supply of 100 000 A at a voltage of 6V. Calculate how many kiliowatt hours (kW h) it takes to make 1 kg of aluminium.
b) At 15p per kW h, what would be the cost of the electrical energy needed to extract 1 kg of Al?
c) Write the overall equation for the electrolysis of Al_2O_3 and hence calculate the mass of carbon anode oxidized to carbon dioxide for every 1 kg of aluminium produced. The reaction at the anode is shown in eqn 11.2.

$$C_{(s)} + O_{2(g)} \rightarrow CO_{2(g)} \tag{11.2}$$

11.3 CLEANING YOUR SILVERWARE WITH ALUMINIUM FOIL

Tired of polishing the silver? Well, waste no more time. The easiest way to clean silver cutlery is to get a basin of salt water, wrap your silverware loosely in aluminium foil, immerse the foil in the salt water and wait.

Q3.

a) Use the electrode potentials in Table 11.1 to explain how the procedure outlined above cleans the silver.
b) Write a balanced equation for the reaction and calculate the standard cell potential.

Table 11.1 Standard electrode potentials for $Al^{3+}{}_{(aq)} + 3e^- \rightarrow Al_{(s)}$ and $Ag^+{}_{(aq)} + e^- \rightarrow Ag_{(s)}$.

Metal half cell	*Standard electrode potential/eV*
$Al^{3+}{}_{(aq)} + 3e^- \rightarrow Al_{(s)}$	−1.66
$Ag^+{}_{(aq)} + e^- \rightarrow Ag_{(s)}$	+0.80

11.4 ALUMS

Aluminium may be a relatively recent addition to the chemist's and engineer's palate, but alums, from which aluminium gets its name, have recorded uses as far back as 1000 BCE. Although, technically, some alums contain no aluminium at all, the general formula for an alum is $MM'(SO_4)_2 \cdot 12H_2O$. In this formula M = a group I metal or NH_4^+ and M' = a trivalent cation, frequently but not always Al^{3+}. Alums were used as mordants, which are chemicals used to help fix dyes to cloth. Indeed, different alums can modify the colour of the same dye quite considerably. Alums are very soluble in water and form acidic solutions (eqn 11.3 and 11.4).

$$KAl(SO_4)_2 \cdot 12H_2O_{(s)} \rightarrow K^+{}_{(aq)} + Al^{3+}{}_{(aq)} + 2SO_4{}^{2-}{}_{(aq)} + 12H_2O_{(l)} \qquad (11.3)$$

$$[Al(H_2O)_6]^{3+} \rightleftharpoons [Al(H_2O)_5OH]^{2+} + H^+{}_{(aq)} \quad K_a = 1.0\times10^{-5}\ \text{mol dm}^{-3} \qquad (11.4)$$

Q4. Using eqn 11.3 and 11.4 and the K_a value provided, calculate the approximate pH of 237 g dm^{-3} solution, stating any assumptions made.

By the Middle Ages, alums were big business and, at one point, the Papacy had something of a monopoly on the stuff. Later on, it was the alchemists who found ways of making their own alums from local clays, proving that, when not attempting to convert lead into gold, they did make some genuine contributions to chemistry and commerce.

11.5 MOLECULES WITH ALUMINIUM

From its position in the periodic table, we might expect aluminium to have a slightly schizophrenic nature; to the left magnesium, most definitely a metal, and to the right, silicon. We have seen that aluminium is extracted by electrolysis and so Al^{3+} ions must be involved but, when we get to other compounds of aluminium, such as aluminium chloride, things get a bit more interesting. For a start, aluminium chloride sublimes at the comparatively low temperature of 178 °C. Compare this with sodium chloride, which melts at 801 °C. It turns out that aluminium chloride ($AlCl_3$) is a molecular species in the vapour state and, when it re-sublimes, it forms a molecular dimer Al_2Cl_6.

Q5.

a) Draw the molecular shape of $AlCl_3$.
b) Explain what is meant by the term "Lewis acid" and why $AlCl_3$ can be categorized as such.
c) Explain how Al_2Cl_6 forms and draw its structure showing all bond angles.

11.6 ALUMINIUM AND ITS ALLOYS

Aluminium continues to perplex when we consider that, despite its position in the reactivity series (well above iron), it is very resistant to corrosion. The apparent contradiction is resolved when the nature of the oxide layer that coats the surface of aluminium is understood. Unlike the flaking rust that forms on iron, the aluminium oxide layer seals itself off from further oxidation. Not only does this prevent the aluminium from further oxidation, but the layer can be thickened in a process called anodizing, in which the aluminium is deliberately connected up as an anode and the surface is oxidized to provide a layer up to about 300 μm thick. During the process, colours can be added and the resultant tints are far more enduring than a coat of paint.

It is rare that pure metals have the properties that enable them to be used without being alloyed with other elements. Aluminium is no exception and, as a pure element, is too soft for practical use. Alloyed with between 1–2% copper, however, it becomes harder and the corresponding material, Duralumin, was patented in 1909 and used extensively in World War I in the manufacture of Zeppelins. A measure of how significant aluminium alloys were to prove is epitomized by the USS *United States*, a ship launched in 1952 that, in containing over 2000 tonnes of aluminium in its super structure, was to lead to a fuel saving of approximately $280 000 per annum.

11.7 THE THERMITE REACTION

There's no doubt that it was iron that made the Industrial Revolution what it was and the railways opened up parts of the world that had previously been inaccessible to all but the hardiest of travelers. However, aluminium was to make an important contribution. The iron rail tracks needed to be welded and for that you needed molten iron. The problem is that temperatures of over 1538 °C are needed to melt iron. If you are laying track across the Rockies in the USA or across Siberia in Russia, this proves something of a logistical

Table 11.2 Standard enthalpy of formation for Al_2O_3 and Fe_2O_3.

Compound	*Standard enthalpy of formation/kJ mol^{-1}*
Al_2O_3	−1675.7
Fe_2O_3	−824.2

problem; no camp fire is going to melt iron! The solution came in the form of the thermite reaction the equation for which is shown in eqn 11.5.

$$2Al_{(s)} + Fe_2O_{3(s)} \rightarrow Al_2O_{3(s)} + 2Fe_{(l)} \quad (11.5)$$

Q6.

a) Use the standard enthalpy of formation data in Table 11.2 to calculate the enthalpy change for the reaction in eqn 11.5.
b) Let's say 2.5 kg of iron are needed to weld two tracks together. Calculate the total mass of both reactants needed to give this quantity of iron.

11.8 IS ALUMINIUM HARMFUL?

We have already acknowledged that aluminium has no known biological role but another question might be does it have any negative impact on biological systems? The answer, in some contexts, seems to be yes and, given its abundance in the Earth's soils, an understanding of this impact is important. It has been claimed that aluminium is the most significant growth-limiting factor in acidic soils across the world. Many important crops depend upon an external source of nitrogen to synthesize proteins. Any decrease in the ability of a plant to absorb soluble forms of nitrogen from the soil will have an impact on the protein content of the crop. It has been shown that aluminium ions tend to bind to negatively charged residues in clay soils, and thus are not readily absorbed by plants. However, if a soil pH drops below 5.5–6.0, then protons begin to exchange with these aluminium ions, which are released into the immediate region around the roots called the the rhizosphere. There is some evidence that this can have a negative impact on the absorbance of nitrate ions and, consequently, the plant protein content. The other impact is on the plant cell aluminium ion concentration, which can rise significantly in acidic soil.

So, if important foodstuffs absorb more aluminium in acidic soil, is there any danger to us? There has been a good deal of controversy about aluminium and Alzheimer's disease but a conclusive connection between the two has yet to be made. Many autopsies on the brains of patients who had Alzheimer's showed raised aluminium levels; however, it was later shown that the aluminium was likely to have come from the autopsy procedure itself.

11.9 CONCLUSIONS

For a biochemist, aluminium is a bit of a disappointment; there's really not a lot to say. For engineers, however, aluminium has been one of the most important materials in usage since it was first isolated in 1825. Today, it is second only to iron in terms of the quantity used as a metal. In 2010, 40.7 million tonnes of new aluminium was extracted but, as it is considerably more expensive than iron to extract, a greater emphasis on recycling is likely to be an important goal in the management of this resource in the future.

Answers to questions in this chapter are on pages 201–203.

CHAPTER 12

Hydrogen

"In the beginning, there was hydrogen...", well, OK, not right at the beginning. Before the inconceivably short time of 1.33×10^{-43} seconds "nothing scientific can be said about the conditions" of the universe.[1] For the next eight minutes or so, protons, electrons and neutrons formed and consorted to form matter that was about 75% hydrogen and 25% helium. Today, those proportions have not changed much and less than 1% of the observable mass of the universe constitutes the rest of the periodic table. Down here on Earth we have a hugely disproportionate amount of the heavier elements and most elemental hydrogen around when the Earth formed 4.6 billion years ago has long since drifted off into space. However, chemically bonded hydrogen is very much still with us as water and numerous other compounds, and our planet is lucky enough to orbit the right kind of sun, at the right kind of distance, to allow water to exist in all three states at the planet's surface.

In this chapter we are going to consider what is probably one of the most common molecular species in the universe: water. Hydrogen does, after all, mean "**water generator**" and so it is in this chapter that this remarkable molecule will be considered. We will also get the chance to look at pH, a little bit of proton NMR and the all important hydrogen bond. Even though it makes a relatively minor contribution in terms of mass, it is, in terms of number, the tenth most common element at the Earth's surface.

Around the World in 18 Elements
By David A. Scott

Published by the Royal Society of Chemistry, www.rsc.org

12.1 HYDROGEN: A UNIQUE ELEMENT

When I started chemistry at school in the 1970s, hydrogen's position in the periodic table was firmly at the top of group I. In terms of its electronic structure this is a perfectly reasonable place to put it. Its solitary electron in its one and only shell means that it sits nicely above Li (2.1) Na, (2.8.1) and so on. However, in many ways it is very unlike the other group I elements. It is a gas at room temperature and forms diatomic molecules, like the halogens in group VII. So, further consideration suggests a position at the top of group VII might be more appropriate. Hydrogen may only have one electron in its shell but, as the first shell can contain only two electrons, it is also one electron short of a full shell, *i.e.*, iso-electronic with helium. Gradually, periodic tables started appearing with hydrogen in both positions, which inevitably led to even more confusion. The general consensus these days is to give hydrogen its own group reflecting its unique properties. It is the only element that has no neutrons (although there are isotopes of hydrogen that do) and its role in bonding is vital in every sense of the word.

Q1.

a) Hydrogen is a solid below -259 °C. Lithium is a solid below $+181$ °C. Describe the bonding in both solids and explain the large difference between the two melting points.
b) There is some evidence that, under extreme pressures, hydrogen might behave as a metal. Suggest a bonding model that might be in agreement with this evidence.

12.2 SOLVING THE SUN'S ENERGY PROBLEM

William Thomson, 1st Baron Kelvin (or Lord Kelvin for short) was, like many scientists of his generation, somewhat mystified as to where the Sun's energy came from. The idea that it was something akin to a vast combustion reaction was unsupportable because a) there was no oxygen in space and b) even if there was, the Sun, as large as it is, would have "run out" long ago. Kelvin suggested that, somehow, the gradual collapsing of the Sun under its own gravitational energy would result in the energy being converted to heat and light. However, again, using the classical mechanical and thermodynamic models of the time, the Sun's mass would only account for about 25 million years of energy. Geological evidence was pointing to an Earth age over 180 times that value. The contradiction would require the formulation of an entirely new type of physics and the solution to the problem is to be found in the world's most famous equation:

$$E = mc^2 \tag{12.1}$$

This equation can cause chemists to blush and start mumbling because it flies in the face of one of the key principles of chemistry: that of *the conservation of mass,* which holds for all **chemical** reactions. $E = mc^2$, put quite simply, means that mass can be converted directly into energy. The symbol c represents the speed of light (3×10^8 m s^{-1}), so c^2 is 9×10^{16}, which is a big number. This means that a very small mass can be converted into a very large amount of energy. Luckily, the activation energy for this reaction is so high that it does not happen at temperatures lower than the interior of the Sun.

Q2.

a) Our Sun converts about 1.5×10^{13} kg of mass into energy per hour. How much energy does this equate to?
b) Using your answer to part a) calculate the Sun's power rating.
c) Show that the joule (J) can be derived from mass (kg) and speed squared (m s^{-1})^2. NB. You might like to use a unit called the yotta joule (YJ); 1 yotta joule $= 10^{24}$ kJ.

As is so often the case, the solution to one problem helps solve others. The reactions in which hydrogen nuclei smash together to convert mass into energy also results in the formation of new heavier nuclei, which, in chemists' terms, means the rest of the periodic table. Stars not only generate energy but also make elements!

12.3 HYDROGEN PRODUCTION AND USAGE

Current estimations of hydrogen production worldwide stand at 3.5×10^{11} m^3 and about half of this is used for the manufacture of ammonia in the Haber process (see Chapter 3 on nitrogen). There are many sources of hydrogen but, at present, the main industrial method of preparing it is by the steam reformation of methane. This takes place in two steps (eqn 12.2 and 12.3):

1. $$CH_{4(g)} + H_2O_{(g)} \rightarrow CO_{(g)} + 3H_{2(g)} \ \_\_\_\_\_\_ \text{ kJ mol}^{-1} \tag{12.2}$$

2. $$CO_{(g)} + H_2O \rightarrow CO_{2(g)} + H_{2(g)} \Delta H^\theta{}_R = -41 \text{ kJ mol}^{-1} \tag{12.3}$$

Q3.

a) In principle any gaseous hydrocarbon can be reformed to make carbon monoxide and hydrogen. Complete the following algebraic expression for the reformation of hydrocarbon C_XH_Y using X and Y as stoichiometric coefficients.

$$C_XH_Y + \_\_\_\_ H_2O \rightarrow \_\_\_\_ CO + \_\_\_\_ H_2$$

b) Now write a balanced equation for the steam reformation of butane.

In the following questions we are going to consider the thermodynamics of this important two-step process.

Q4.

a) Use the data in Table 12.1 to draw a Hess's cycle and calculate a value for the enthalpy change in step 1 (eqn 12.2).
b) Comment on the choice of a temperature of 1100 K for this step in industry.
c) How would the temperature of 1100 K affect the second step of the reaction? Explain your answer.
d) Suggest how conditions could be altered to favour the forward reaction in **both** steps simultaneously.

In previous chapters we have already come across an equation that links a great deal of physical chemistry and I am happy to re-state it here again in eqn 12.4:

$$\Delta G = \Delta H - T\Delta S = -RT\ln K = -nFE \tag{12.4}$$

For our purposes we are going to extract:

$$\Delta H - T\Delta S = -RT\ln K \tag{12.5}$$

Re-arranging and cancelling, we obtain eqn 12.6 and then 12.7:

$$\ln K = -\Delta H/RT - (-T\Delta S/RT), \tag{12.6}$$

$$\text{so } \ln K = -\Delta H/RT + \Delta S/R \tag{12.7}$$

Table 12.1 ΔH^{θ}_{f} values for the species in eqn 12.2.

Species	ΔH^{θ}_{f}/kJ mol^{-1}
CH_4	−74.8
H_2O	−241.8
CO	−110.5
CO_2	−393.5
H_2	0

As $\Delta S/R$ is a **temperature independent** constant, we can now write eqn 12.8:

$$\ln K = -\Delta H/RT + c \quad (12.8)$$

Given the enthalpy change for a reaction, we can make a quantitative prediction to the effect of temperature change on the equilibrium constant.

Let's see if we can make some predictions about the thermodynamics of the second step.

$$CO_{(g)} + H_2O \rightarrow CO_{2(g)} + H_{2(g)} \; \Delta H^{\theta}{}_{R} = -41 \text{ kJ mol}^{-1} \quad (12.9)$$

Q5.

a) Write an expression for K_P for the above reaction (eqn 12.9) and explain why the constant has no units.
b) The value of K_P for the above reaction in eqn 12.9 at 700 K is 8.13. Calculate a value for K_P at 1100 K.
c) Does this quantitative prediction agree with the qualitative prediction you would make on the basis of Le Chatelier's principle? Explain your answer.
d) What conditions other than the position of equilibrium need to be considered in the industrial manufacture of a product?

12.4 WATER AND THE DEVELOPMENT OF CHEMISTRY

In the 2500 years since Thales of Miletus (624–546 BCE) suggested that all matter was a form of water our understanding of this remarkable molecule has evolved. After Empedocles (490–430 BCE) developed the idea further into four basic elements, earth, air, fire and water, progress rather stalled until experiments in the 18th century showed that water could be decomposed into something simpler (so it couldn't, by definition, be an element!). Henry Cavendish showed that "flammable air" (hydrogen as we would now call it), when combusted, produced water. The water was produced by the combination of flammable air with "dephlogisticated" air, which reads very strangely to a modern chemist. This is due to the fact that the concept of oxygen as an element had to wait until Lavoisier (but more on this in Chapter 17 on oxygen). Even after the idea of water as a compound of hydrogen and oxygen was accepted, John Dalton still believed that the molecular species consisted of one hydrogen atom bonded to one oxygen atom. Dalton intuitively felt that the simplest

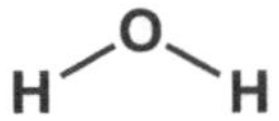

Figure 12.1 Structure of water.

combination of the two elements made most sense. In 1866 August Wilhelm von Hofmann threw a spanner in the works when he showed that the electrolysis of water produced about twice the volume of hydrogen to oxygen. (Actually a little more than twice the volume as oxygen is slightly soluble in water and hydrogen is not.) Today, X-ray crystallography supports bonding theory: water is H_2O. But more than that: it adopts a V shape with an average bond angle of 104.5° and has a dipole as a result of differences in electronegativity between oxygen and hydrogen and its shape (Figure 12.1).

Q6.

a) Explain in detail why a water molecule has a "V" shape with an average bond angle of about 104.5°.
b) Explain why a water molecule, with a relative atomic mass of 18 g mol^{-1}, has such an unexpectedly high boiling point (100 °C; *c.f.* SO_2, with a mass of 64 g mol^{-1}, has a melting point of 76 °C).

Other remarkable properties of water include:

1. Its ability to dissolve a large range of substances from ionic compounds and small molecular species, *e.g.*, O_2 and CO_2, and larger molecules, such as sucrose and amino acids.
2. Its relatively high specific heat capacity, which enables it to absorb and retain heat energy and release it slowly.
3. The fact that, when frozen, it is less dense than liquid water, and thus ice floats in water.

Q7. "If water was a linear molecule, life could never have evolved". Discuss this statement. (Now you don't get many questions like that in chemistry!)

12.5 WATER AND THE pH SCALE

Having established that water is molecular H_2O and that many of its special properties are related to hydrogen bonding, we might be forgiven for expecting water to behave as a molecular solution. However, water still has a surprise for us. The one thing that every primary school pupil knows (hopefully) is that you never put water on an electric fire. Why? Well because water conducts electricity of course. But hold on, a molecular liquid shouldn't conduct electricity should it? The first suspicion must be that water contains lots of things dissolved in it, ionic salts for example. So removing those should reduce the conductivity of water, which it does **but** not to the extent that we would expect. No matter how much you purify water, it still behaves like there are ions present to carry an electric charge. Conclusion? Maybe water itself forms ions! And it does (eqn 12.10).

$$H_2O \rightleftharpoons H^+_{(aq)} + OH^-_{(aq)} \tag{12.10}$$

But pure water is actually a very poor conductor of electricity, which tells us that the equilibrium shown in eqn 12.10 lies on the left hand side to a degree that we can quantify by writing an equilibrium expression for the dissociation (eqn 12.11):

$$K_C = [H_3O^+][OH^-]/[H_2O] = 1.81\times10^{-16}\ \text{mol dm}^{-3} \tag{12.11}$$

Now, we can see that such a small number means that the water is hardly ionized at all. So, we can make an approximation. If we have 1000 g of water, which is equivalent to 1000/18 = 55.5 moles of water. Now, if water does ionize then some of these 55.5 moles must dissociate **but**, in fact, so few actually do dissociate that we can assume that the number of water molecules remains at 55.5. If we do this, then we can consider the denominator in the K_c expression (eqn 12.11) to stay at 55.5. Rearranging our equation gives us eqn 12.12:

$$(1.81\times10^{-16})\times 55.5 = [H_3O^+][OH^-] = 1.00\times10^{-14}\ \text{mol}^2\ \text{dm}^{-6} \tag{12.12}$$

We how have a special quantity for water called the ionic product (symbol: K_w), which is constant for all aqueous solutions **at 25 °C.** Like all equilibria, this value is dependent upon temperature.

There's nothing like a bit of cognitive conflict to wake students up. The one thing they all know is that a pH of 7.0 means a neutral solution so, when I tell them that at 60 °C pure water is still neutral but has a pH of about 6.6, they start complaining about how they were lied to at GCSE...again! I try to reel things in by saying that, at the end of the session, they will appreciate that there is no conflict.

We have seen above that K_w for water at 25 °C = 1.00×10^{-14} mol^2 dm^{-6}. But at 60 °C it isn't anymore! At 60 °C, $K_w = 5.6\times10^{-14}$ mol^2 dm^{-6}.

Q8.

a) Use the value for K_w at 60 °C to show that pure water at this temperature has a pH of 6.6.
b) The pH value decreases as the temperature rises. What does this tell you about the effect of temperature on the ionization of water?

12.6 HYDROGEN BONDING

We have briefly mentioned hydrogen bonding but it is impossible to overstate its significance in biochemistry and life. There are three elements that have a sufficiently different electronegativity value with hydrogen to enable hydrogen bonds to form. Fluorine is the least interesting by virtue of the fact that the only hydrogen and fluorine species is HF. The other two are nitrogen and oxygen and because of their ability to form three and two covalent bonds, respectively, hydrogen bonding can become a feature of many of the molecules containing these elements. Consider amines, alcohols, amino acids, sugars and many others (Figure 12.2). Perhaps the most important thing about the hydrogen bond is that it is quite strong but not **too** strong! Such a bond allows DNA double helixes to uncoil and replicate and it allows enzymes to form stable structures but not to the extent that they are completely inflexible, which would not allow them to bind to their substrates.

Q9. The two molecules in Figure 12.2 represent the DNA bases adenine and thymine. Redraw the molecules showing how the two molecules form two hydrogen bonds with each other.

12.7 PROTON NMR

There can be few analytical techniques that have had as far reaching impact as nuclear magnetic resonance (NMR) spectroscopy, particularly in the biosciences and the pharmaceutical industry. NMR spectroscopy

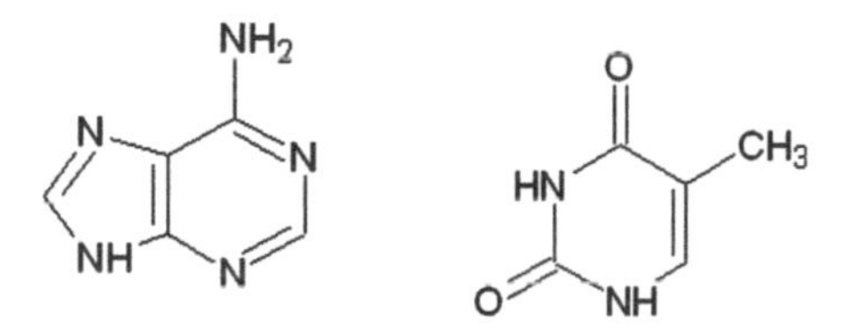

Figure 12.2 Structures of adenine (left) and thymine (right).

is now routinely used to gain structural information about large protein molecules. It has the advantage of X-ray diffraction in that it allows information about the dynamic nature of the protein to be studied. X-Ray crystallography, as the name suggests, requires the development of a crystalline form of a protein and this necessarily renders the protein static.

Read the passage below and then answer the question that follows.

> *"Like all types of spectroscopy, nuclear magnetic resonance spectroscopy (NMR) is all about the interaction between a part of the electromagnetic (EM) spectrum and matter. In this particular case it is at the low energy radio end of the spectrum, where radio waves can interact with a property called nuclear spin. As not all nuclei have the quantum property of spin it only interacts with particular nuclei. The most abundant hydrogen isotope ^{1}H (sometimes called protium) can adopt quantum spin values of $+1/2$ and $-1/2$ and it is the transition from the $+1/2$ to the $-1/2$ spin that first absorbs and then in returning from $-1/2$ to $+1/2$ re-emits a particular frequency of EM radiation. If all hydrogen nuclei re-emitted radiation of the same frequency then NMR's analytical value would be zero but it turns out that there are slight variations in the "resonance" of protons that are indicative of particular proton environments. In addition, NMR spectra give information about the relative abundance of different proton environments and, at high resolution, can give relative positional information due to a process known as spin–spin coupling."*

Q10.

a) Explain what you understand by the term "proton environments."

The skeletal formulae of two organic structures are shown in Figure 12.3.

b) (i) Name both molecules.
(ii) Outline what you would expect to be the main features of the proton NMR spectrum for each molecule. (You would not be expected to know the specific chemical shifts but you should be able to say something about the number of proton environments, the relative number of protons in each environment and any expected splitting patterns.)

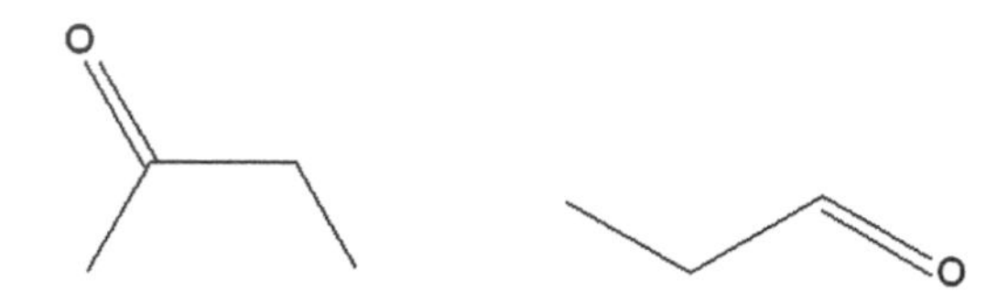

Figure 12.3 Structures of two organic molecules (see Q10).

12.8 CONCLUSIONS

I have only scratched the surface of hydrogen chemistry in this chapter. I have said nothing about acids, hydrogen fuel cells, metal hydrides, the importance of proton gradients for generating energy in cells and probably several hundred other worthy applications involving hydrogen. Hopefully, by mentioning them here, you may feel inspired to look a few up. Have fun!

REFERENCE

1. N. N. Greenwood and A. Earnshaw, *Chemistry of the Elements*, 2005, Elsevier Butterworth Heineman, Oxford, pp. 1–19.

Answers to questions in this chapter are on pages 204–207.

CHAPTER 13

Chlorine

Gas! Gas! Quick, boys! – An ecstasy of fumbling,
Fitting the clumsy helmets just in time;
But someone still was yelling out and stumbling,
And flound'ring like a man in fire or lime...
Dim, through the misty panes and thick green light,
As under a green sea, I saw him drowning.
In all my dreams, before my helpless sight,
He plunges at me, guttering, choking, drowning.

Dulce et Decorum est, verse 2, Wilfred Owen, 1917

Chlorine gas was used as a chemical weapon for the first time in World War I. Although it was not the first gas used in warfare, its effects are etched into the mind of anyone who has ever read Wilfred Owen's *Dulce et Decorum est* even though more deadly gases have since been developed. In fact chlorine's war record is a grim one: phosgene and mustard gas are both chlorine-containing gases used against the troops in World War I (and, all too sadly, other more frequent conflicts, as the Iraqi Marsh Arabs will testify) and, in Vietnam, Agent Orange was used to defoliate the forests to expose the Viet Cong. But, on the flip side, chlorine-containing organic molecules have been used as disinfectants, anticancer drugs and in water

Around the World in 18 Elements
By David A. Scott

Published by the Royal Society of Chemistry, www.rsc.org

treatment. As a chloride ion, chlorine is the most abundant anion in the oceans and it is essential to normal bodily function.

13.1 CHLORINE: A MEMBER OF GROUP VII, THE HALOGENS

Group VII in the periodic table is the only one to have elements that exists as solids, liquids and gases at room temperature. Chlorine is a yellow/greenish gas at room temperature and is instantly recognizable by its smell. The group shows a clear trend in physical properties (Table 13.1).

Q1.

a) Draw a dot and cross diagram of a chlorine molecule showing all of the valence electrons.
b) Explain what is meant by the term *dispersion forces* (also known as London forces).
c) How do dispersion forces account for the trend in boiling points of the group VII elements?

13.2 PREPARATION OF CHLORINE

It was Carl Wilhelm Scheele (1742–1786) who first prepared chlorine gas in Sweden in 1774. On first analysis, his method of preparation may seem implausible. It consisted of reacting manganese IV oxide with hydrochloric acid (eqn 13.1 and 13.2).

$$MnO_2 + 4H^+ + 2e^- \rightleftharpoons Mn^{2+} + 2H_2O \quad E^\theta = +1.23\ \text{eV} \tag{13.1}$$

$$Cl_2 + 2e^- \rightleftharpoons 2Cl^- \quad E^\theta = +1.36\ \text{eV} \tag{13.2}$$

Table 13.1 Boiling points of the group VII elements.

Element	*Boiling point/°C*
F	−188
Cl	−35
Br	+59
I	+184
At	–

Q2.

a) Use the half equations in eqn 13.1 and 13.2 to explain why you might not expect manganese IV oxide to produce chlorine gas in a reaction with hydrochloric acid.

The Nernst equation (eqn 13.3) shows the following relationship:

$$E = E^{\theta} - \frac{RT}{nF} \ln Q, \qquad (13.3)$$

where Q = [reduced species]/[oxidized species] (13.4)

b) Define the standard conditions for making standard electrode potential measurements.
c) Use the Nernst equation (eqn 13.3) to calculate E values for the two half cells for a 6 M solution of HCl. (Note the H^+ ion quantity in the manganese IV oxide half equation.)
d) Using the Nernst equation, show that using 6 M hydrochloric acid and a temperature of 80 °C allows the reaction in part a) to become feasible.

13.3 CHLORINE IN THE PREVENTION OF INFECTION

In wasn't until the 19th century that a clear link between microbes and disease was finally made. Following Louis Pasteur's experiments in the 1860s, the Germ Theory of disease was established and medics began to conduct experiments under aseptic conditions. Joseph Lister's use of carbolic acid spray prior to surgery was to improve post-operative recovery considerably. For the first time, surgery didn't mean almost certain death.

Q3.

a) Carbolic acid is an aqueous solution of phenol. Phenol dissolves in water to form a weakly acidic solution. Draw the skeletal structure of phenol.
b) Draw the skeletal formula of the phenoxide ion and use it to explain what is meant by "a resonance structure".
c) Write down an expression for the dissociation of phenol in water and calculate the pH of a 0.002 M solution of this acid given $K_a = 1.28 \times 10^{-10}$ mol dm^{-3}.
d) An improvement on phenol as an antiseptic was 2,4,6-trichlorophenol (2,4,6-TCP), which proved to be less

corrosive than phenol. Draw out the skeletal structure of this molecule.

e) Suggest a synthetic procedure for converting phenol into 2,4,6-trichlorophenol, explaining your choice of reagents and conditions.

f) Describe what you would expect to be the main features of a proton NMR spectrum of 2,4,6-TCP, giving reasons for your suggestions.

g) The K_a value for 2,4,6-TCP is 1×10^{-6} mol dm^{-3} at 298 K. Describe and explain the impact of the three chlorine atoms on the K_a values by comparing phenol and 2,4,6-TCP.

13.4 CHLORINE AND WATER STERILIZATION

Of all modern technologies, a strong argument can be made that the provision of a clean, safe water supply has saved more lives than any other. Cholera, typhoid and dysentery are amongst those diseases that were once common in cities like London and New York but are now things of the past. Sadly, the same cannot be said for many developing nations, where many of these diseases are still endemic. A simple and effective way to deal with waterborne pathogens is to bubble chlorine gas through water. The chlorine both dissolves in and reacts with the water and the following equilibrium (eqn 13.5) is set up:

$$Cl_2 + H_2O \rightleftharpoons HCl + HOCl \qquad (13.5)$$

The two acids formed are hydrochloric acid and hypochlorous (chloric I acid). It is the second acid that has the antibacterial properties and, as a weak acid in solution, the following equilibrium (eqn 13.6) is set up:

$$HOCl \rightleftharpoons H^+ + ClO^- \qquad (13.6)$$

Although both chlorine-containing species on the left and right hand side of this equilibrium are antibacterial, it is the protonated neutral species on the left hand side that has the more potent antibacterial effect. For this reason, pH regulation is an important factor to consider in ensuring water quality. In swimming pools, for example, the pH has to be a compromise between a sufficiently acidic solution to keep the above equilibrium to the left hand side, but not too acidic so that swimming becomes uncomfortable. Other buffers in swimming pools aim to keep the pH at between 7.4 and 7.6.

Table 13.2 Standard electrode potentials for $2H^+ + S + 2e^- \rightarrow H_2S$ and $Cl_2 + 2e^- \rightarrow 2Cl^-$.

Half equation	*Standard electrode potential/eV*
$2H^+ + S + 2e^- \rightarrow H_2S$	+0.14
$Cl_2 + 2e^- \rightarrow 2Cl^-$	+1.36

Q4.

a) Explain why eqn 13.5 represents a disproportionation reaction.
b) The pK_a value for chloric I acid (eqn 13.6) is 7.4. Write an expression for the K_a of chloric I acid and calculate the K_a value.
c) If the pK_a of chloric I acid = 7.4 and the ideal pH for a swimming pool is also 7.4, how might you go about preparing a swimming pool solution of 7.4?
d) It is believed that many pathogenic bacteria are coated with a negatively charged slime. Suggest why HOCl is likely to be a more effective bactericide than OCl^-.

Some water supplies have a rather pungent odour of rotten eggs caused by dissolved hydrogen sulfide.

Q5. Use the half equations in Table 13.2 to show how chlorine can remove this odour from the water supply.

13.5 CHLORATE I (OR HYPOCHLORITE)

An alternative to using chlorine to sterilize water is to use sodium chlorate I (sometimes called sodium hypochlorite). As a solid, this is easier to transport than chlorine but its storage can be a problem due to its gradual decomposition. Such reactions are one of the reasons why chemists study rates of reaction. An analysis of the reaction rate may give us insights into reaction mechanisms. This, in turn, may suggest the best way to control a reaction rate, in this case to slow down the decomposition. The main decomposition reaction involved can be shown by eqn 13.7.

$$3ClO^- \rightarrow ClO_3^- + 2Cl^- \qquad (13.7)$$

It is important to realize that the stoichiometric equation shows us the ratio of reactants and products but we can deduce nothing about the mechanism of the breakdown from the equation on its own. The only way to get information about the reaction mechanism is to follow the reaction under controlled conditions whilst varying the concentration of the different reactants under controlled temperatures.

Experiments into the decomposition reaction in eqn 13.7 revealed the following rate equation (eqn 13.8):

$$\text{Rate} = k[\text{ClO}^-]^2 \tag{13.8}$$

Q6.

a) What would be the units for the rate constant for this reaction?

A suggested rate mechanism from the rate equation is as follows:

$$1.\ 2\text{ClO}^- \rightarrow \text{ClO}_2{}^- + \text{Cl}^- \tag{13.9}$$

$$2.\ \text{ClO}_2{}^- + \text{ClO}^- \rightarrow \text{ClO}_3{}^- + 2\text{Cl}^- \tag{13.10}$$

b) Which of the two steps is likely to be the rate determining step? Justify your choice.

The Arrhenius equation (eqn 13.11) shows the relationship between rate constant k, the activation energy and the temperature in K:

$$k = A\text{e}^{-E/RT},$$

where k = rate constant, A = pre-exponential constant, E = activation energy, R = the universal gas constant (8.314 J K^{-1} mol^{-1}) and T = temperature in K.

c) Given $k = 11.4$ at $T = 333$ K and $A = 2.1 \times 10^{12}$, calculate the activation energy for this reaction in kJ mol^{-1}.
d) One of the recommendations for sodium chlorate I storage is to dilute the solution to reduce the rate of decomposition. How does the rate mechanism suggest that this might work?

13.6 CHLORINE'S ISOTOPIC SIGNATURE

Naturally occurring chlorine has two stable isotopes, ^{35}Cl and ^{37}Cl, which occur in the approximate ratio of 3 : 1. Any molecules that contain a single chlorine atom will give two molecular ion peaks, two units apart. Such ratios can be effective signatures for identifying elements in simple organics molecules.

Q7.

a) Analysis of a chlorohydrocarbon shows it to have the following composition by mass: 24.24% carbon, 4.04% hydrogen and 71.71% chlorine. Calculate the empirical formula of this molecule.
b) Further analysis of this molecule showed three molecular ion peaks, two mass units apart from each other in a ratio of 9 : 6 : 1. What can we deduce about the molecule from this?
c) Suggest a molecular formula for the molecule described in part a).

13.7 CHLORINE-BASED MEDICINES AND THE TREATMENT OF CANCER

Melphalan (Figure 13.1) is an early example of a chlorine-containing anticancer drug and was developed in the UK in 1954.

It was something of an improvement on some of the earlier drugs because it was more soluble in water, and thus could be administered orally.

Q8.

a) Melphalan has the ability to form a zwitterion in aqueous solution. Redraw the molecular zwitterion and explain why this form of the molecule is likely to be water soluble.
b) Draw the structural formula of the molecule produced when melphalan is heated under reflux with propan-2-ol.

13.8 SOME MORE CHLORINE-CONTAINING MOLECULES

Chlorine shows considerable diversity in its ability to form bonds. As Cl^-, it is the most common ion in the sea. It forms stable bonds with carbon and exhibits oxidation states of -1, $+1$, $+3$ and $+5$, as well

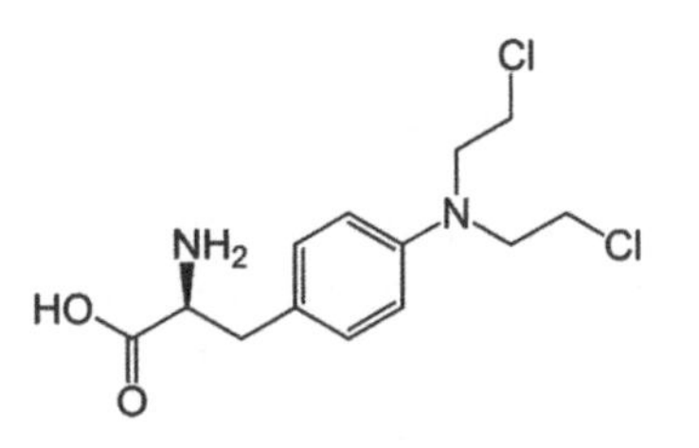

Figure 13.1 Structure of melphalan.

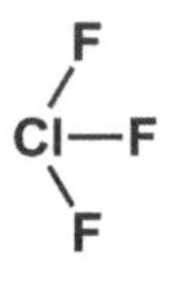

Figure 13.2 Structure of ClF_3.

as a few molecules like ClO_2, which bonding theory finds difficult to account for. One interesting molecule is ClF_3 (Figure 13.2). In this molecule, chlorine opens up the 3d subshell and is in oxidation state +3. As a result, there are five pairs of electrons in the valence shell, three bonding and two non-bonding. The shape that immediately suggests itself is a trigonal planar structure with all three Cl–F bonds in the plane and the lone pairs above and below at 180° from each other. However, in reality, the molecule adopts a distorted "T" shape, which has a lower energy configuration. This may initially seem counterintuitive until we remember that a) lone pairs have a greater repulsive effect than bonding pairs and b) a T shape can be distorted to reduce repulsive effects, whereas no distortion is possible in a trigonal planar structure.

Q9.

a) Draw electrons in boxes to show the sp^3d hybridization in ClF_3.
b) Draw the shape of the molecule ClF_5, describing the level of chlorine hybridization in the molecule and marking on all bond angles.

13.9 CHLORIDE IONS AND THE CELL

You are provided with the following data about chloride ions and living cells.

Intracellular chloride ion concentration = 4–5 mmol dm^{-3}.

Extracellular chloride ion concentration = 113–116 mmol dm^{-3}.

What do you conclude?

Well, it is clear that the distribution of chloride ions inside and outside the cell is far from equilibrium and, as this a general feature of all living cells, there must be cellular processes that work very hard to keep the system away from equilibrium (see Chapter 5 on calcium). The importance of this dis-equilibrium for normal cell functionality can be seen when it goes wrong. Cystic fibrosis is a disease that is associated with a mutation on chromosome 7, which results in the malfunctioning of a transmembrane chloride ion regulator. Such mutations give rise to many of the extremely unpleasant traits associated with cystic fibrosis, such as the build up of mucus in the lungs and an impaired ability to digest food.

Q10.

a) The average human blood volume is 5 dm^3, of which 60% is plasma. Plasma is an extracellular fluid. Using the data above, calculate the range for the total mass of chloride ions dissolved in this volume of plasma.
b) If the average amount of chloride in the human body is 95 g, calculate the percentage of chloride that is dissolved in blood plasma.

13.10 CFCs

When it comes to chlorine compounds, the elephants in the room, so to speak, are the chloro–fluoro carbons (CFCs). Their impact on the ozone layer was first suggested in 1974 by Rowland and Molina and, by 1985, the British Antarctic Survey had found evidence that the ozone layer was indeed being depleted. However, in 1987, 70 nations signed the Montreal Protocol on Substances that Deplete the Ozone Layer, regulating the usage of CFCs, and the way the international community has (largely) phased out CFC usage since represents one of the more notable achievements of international politics and, hopefully, sets a president for future co-operation.

A consideration of the impact of CFCs on the steady state equilibrium between dioxygen and ozone is worth considering. In the upper atmosphere an equilibrium between dioxygen and ozone is maintained in a particular position due to ultraviolet light (13.11).

$$O_3 \rightleftharpoons O_2 + O \tag{13.11}$$

The generation of ozone molecules is vital as it is these molecules that are responsible for absorbing ultraviolet radiation in the 210–230 nm range. This frequency of radiation, if it reached the Earth's surface, would break covalent bonds in water molecules, hydrocarbons and many other molecules. In short it would be devastating for life.

CFC molecules are relatively inert at sea level but, as they rise in the atmosphere, they are increasingly exposed to more energetic radiation. This can result in the homolytic fission of CFC molecules to generate chlorine radicals:

$$CF_2Cl_2 \rightarrow CF_2Cl^{\bullet} + Cl^{\bullet} \tag{13.12}$$

Once formed, the chlorine radical reacts with ozone to form $ClO^{\bullet}$ and O_2. Once formed, the $ClO^{\bullet}$ reacts with an oxygen atom to form O_2 and regenerate the chlorine radical. It is this catalytic decomposition that is so damaging. A single radical may be responsible for breaking down thousands of ozone molecules before any possible termination step.

Q11.

a) Draw the displayed formula of CF_2Cl_2 and show, using appropriate notation, how a chlorine radical is formed when exposed to ultraviolet light.
b) Using equations, show how the chlorine radical is regenerated during ozone breakdown.
c) Suggest why fluorine radicals are not formed when exposed to ultraviolet light.
d) Suggest why the reaction of ClO and NO_2 can be considered to be a termination step.

13.11 CONCLUSIONS

Although organic chlorine compounds do exists in nature, it is widely accepted that the majority are anthropogenic in origin and that their general inert character means that they can accumulate in nature, especially at the top of food chains, where their long term impact may be harmful. Not for the first time, a possible solution to this lies in another part of the food chain. Strains of anaerobic bacteria are now being developed to breakdown compounds, like polychlorinated bisphenols (PCBs), and return the elements to the environment. But there will be those who think that the solution lies less with genetically modified bacteria and more with a significant reduction in consumerism in general.

Answers to questions in this chapter are on pages 208–213.

CHAPTER 14

Zinc

According to the United Nations (UN) population fund, the Earth's human population reached seven billion around the end of October 2011. Some predictions suggest that this figure will reach between nine and ten billion before it begins to level off in the middle of the 21st century. Quite how these millions will be fed is a question open to debate. Some say that the meat-protein rich diet of the US and Western Europe is unsustainable and yet the desire for such a diet is now increasing in the East. Lest we forget, meat protein requires plant protein, and thus an improvement in crop yield is essential. However, growing the necessary crops is only one of the issues. A research review in the journal *The New Phytolygist*[1] has suggested that two thirds of the world's population lack one or more essential mineral elements, the seven named elements being: iron, zinc, copper, calcium, magnesium, selenium and iodine. The need for iron and calcium has been known for some time but it is only relatively recently that the necessity for other elements has become identified. In the case of zinc, the relatively modest amount present—between 2 and 3 g for a healthy adult—masks the vital role that it plays in healthy human development. As is frequently the case, it is only when it is absent that its necessity is fully appreciated.

Dating zinc's discovery is difficult as the concept of a pure element post-dates its early usage. There is evidence that the Romans were making brass (a zinc and copper alloy) at the time of the emperor Augustus (63 BCE–14 CE) and it was almost certainly smelted in India

Around the World in 18 Elements
By David A. Scott

Published by the Royal Society of Chemistry, www.rsc.org

before that. Zinc played a key role in the development of the first electrochemical cell in 1800 and, today, over five million tonnes per year is used for galvanizing alone.

14.1 ZINC: 'd' BLOCK YES, TRANSITION METAL NO

Zinc is at the top of a group in the periodic table that, at the time of writing, has the newest named element in the periodic table. Group IIB (or 12) now consists of Zn, Cd, Hg and Cn (copernicium). Of the four, zinc is the only one you are likely to handle. Cadmium and mercury are both highly toxic and copernicium atoms are unlikely to be stable long enough to have any meaningful chemistry at all. Zinc is the only one of the four that is essential to life.

Q1.

a) Give the s,p,d electronic configuration of Zn (atomic number 30) and Zn^{2+}.
b) Explain why zinc is not considered to be a transition metal even though it is in the 'd' block of the periodic table.
c) Draw the shape of $[Zn(H_2O)_6]^{2+}$ showing all the bond angles and explain why it is colourless in solution.

14.2 ZINC ISOTOPES

Isotopes are, by definition, atoms of the same element but with different mass numbers. This means that their chemistry is essentially the same but, increasingly, it has been noted that biochemical and biogeochemical processes can show distinctive isotopic ratios. Altered ratios for lighter elements, such as carbon and oxygen, are now regularly used to get insight into various physiochemical processes: photosynthesis, for example, has a slight preference for fixing carbon – 12 atoms over the heavier carbon – 13. There is now evidence that some plants have a slight preference for the heavier zinc isotope. Weis *et al.*[2] has presented research that some plants have a selective preference for absorbing the heavier ^{66}Zn over the ^{64}Zn isotope from a nutrient growth solution. The reasons for such a preference remain, in the words of the authors, "speculative" but, if plants are to be engineered in the future to select and concentrate essential minerals from the environment, then a clearer understanding of such processes is essential.

Q2. Use the data in Table 14.1 to calculate the relative atomic mass of zinc to the correct number of decimal places.

Table 14.1 The % abundance of the isotopes of zinc.

Zinc isotopic mass	64	66	67	68
% Abundance	48.75	27.9	4.2	19.15

14.3 ZINC: ABUNDANCE AND EXTRACTION

At 75 ppm, zinc is rather more abundant in the Earth's crust than copper. Extracting zinc from its ore, however, presents a number of challenges. Firstly, zinc has a comparatively low boiling point of 907 °C. During smelting processes, such as those required to extract copper (melting point 1083 °C), zinc simply vaporizes and is immediately oxidized. The zinc vapour needs to be condensed and there is evidence that this was being done in the north west of the Indian sub-continent many hundreds of years before it was finally isolated and named by Andreas Sigismund Marggraf in 1746 in Germany.

A slight philosophical interlude

Is it right to credit Marggraf with the "discovery" of zinc?

Zinc was clearly being used hundreds of years before it was finally considered to be an element. From the Greek point of view, zinc could never have been granted elemental status because all matter was conceived to be made from the four "elements", earth, fire, water and air. An analogous question might be: was the discoverer of fire the first individual to use it way back in pre-history, or was it the first person who defined it as the process of combustion? Even *facts* in science are embedded in conceptual frameworks. The American philosopher W. V. O. Quine noted, observation is not neutral but "theory laden". Is it just a case of saying we see the same stuff but in a different way?

I promise, no more "philosophical interludes"!

14.3.1 Extraction

At present, ores with less than 5% zinc by mass are not considered economically viable to work. "Economical" zinc ore is crushed and separated from other bulk impurities to get a concentrate equivalent to about 55% zinc by mass. At this point, the main zinc ore is zinc sulfide, together with various iron, copper, lead and silver impurities. The zinc sulfide is first converted to zinc oxide in a process called roasting.

Table 14.2 Standard electrode potentials for the extraction of zinc.

Half equation	*Standard electrode potential/eV*
$Zn^{2+}(aq) + 2e^- \rightarrow Zn(s)$	−0.76
$Cu^{2+}(aq) + 2e^- \rightarrow Cu(s)$	+0.34
$2H^+(aq) + 2e^- \rightarrow H_2(g)$	0.00
$2Ag^+(aq) + 2e^- \rightarrow 2Ag(g)$	+0.80
$O_3 + 2H^+(aq) + 2e^- \rightarrow O_2(g) + H_2O$	+2.07
$Fe^{2+}(aq) + 2e^- \rightarrow Fe(s)$	−0.44

After that, in the hydrometallurgical process, sulfuric acid leaches through the impure oxide to produce soluble zinc sulfate. Insoluble sulfates, such as silver and lead, stay behind. In the final purification process zinc powder is added to the solution, which displaces soluble impurities such as copper. The pure zinc sulfate is then electrolyzed.

Q3.

a) Write a balanced equation for the reaction of zinc sulfide with oxygen to produce zinc oxide and sulfur dioxide.
b) If a 10 tonne sample of zinc ore concentrate, which is 88% zinc sulfide, reacts completely with oxygen, what mass of sulfur dioxide is produced?
c) Write a balanced equation, including state symbols, to show the reaction of sulfuric acid with zinc oxide.
d) Use the standard electrode potentials in Table 14.2 to explain how soluble copper impurities are removed from the zinc sulfate solution by the addition of zinc powder.

14.4 ZINC AND THE VOLTAIC PILE

Fifty four years after Marggraf had identified the element zinc, Alessandro Volta (1754–1827) was to build a device that would have a massive impact on science. Throughout the 18th century, physicists had become increasingly fascinated with the phenomenon of electricity and in the first year of the 19th century Volta built a device that was to become known as the voltaic pile. It consisted of alternating discs of zinc and copper separated by cloth soaked in salt water. When connected up, the pile generated a continuous supply of electricity and was to be used in numerous experiments, including the ones that allowed Humphrey Davy and his associates to isolate no less than five new elements.

Q4.

a) Under standard conditions, what potential difference would be generated by connecting a zinc half cell to a copper half cell? (Note: refer to Table 14.2 to help answer this question.)
b) Give **two** reasons why the potential differences generated by Alessandro Volta's cell would be different from that measured in part a).
c) Use the half cell equations in Table 14.2 to explain why there would be a build up of gas at the copper electrode as the voltaic cell was discharged and identify the gas.

14.4.1 Humphrey Davy and the Voltaic Pile

It seems that, in some respects, Humphrey Davy was the Jeremy Clarkson of his day. Having seen, and been impressed by Volta's pile, he decided to build a *really* big one consisting of 2000 cells, which must have generated in excess of 1000 volts. He then went about attempting to separate substances that had previously been thought to be elements in themselves. Two of the substances he successfully decomposed were lime (calcium oxide) and potash (potassium hydroxide).

Q5.

a) Write anode and cathode equations for the electrolysis of i) molten calcium oxide and ii) molten potassium hydroxide.
b) Explain why both compounds must be molten in order to conduct an electric current.

The use of the voltaic pile had a huge impact on chemistry. At the end of the 18th century, thirty two elements were known. In the next eight years a further fifteen would be added to the list. In 1812 Davy said:

> *"Nothing tends so much to the advancement of knowledge as the application of a new instrument."*

By 1869, about 63 elements were known; enough to give Dmitri Mendeleev sufficient clues to propose his periodic table of the elements. Davy, sadly, lived only until the age of 50. He died in Switzerland in 1829 of heart disease exacerbated, no doubt, by his tendency to smell and experiment with many of the chemicals he handled. Davy was also an aspiring poet and his contemporary, Robert Southey (also a poet), almost certainly making a rather barbed pun was supposed to have said about Davy:

> *"He had all the elements of a poet, he only wanted the art."*

14.4.2 The Voltaic Pile and Its Impact on Neuroscience

The voltaic pile was also to give a great insight into the functioning of nerve cells in animals. Luigi Galvani, a contemporary of Volta's, had shown that frogs' legs twitch when connected to two different metals and these observations were taken a step further when Giovanni Aldini used recently hanged felons from Newgate Prison in London. Aldini noted that the lifeless corpses became ridged and sat up when a voltage was applied. This made for compulsive viewing at the Royal Society and it is almost certain that one Mary Wollstonecraft Godwin was present at one of these public demonstrations. Later on, as Mary Shelley, she would write *Frankenstein.* The application of Davy's "new instrument" seems to have had an equally impressive impact on the arts, contributing to the writing of what many consider to be the first science fiction novel.

14.5 THE DANIELL CELL

In 1836 J. F. Daniell made the first improvement to the voltaic pile by making two modifications. Firstly, the copper cathode was now immersed in a solution of copper sulfate and, secondly, a porous clay pot was used to keep the zinc electrode immersed in a sulfuric acid solution, separated in such a way as to allow the sulfate ions to move through the pot but not the metal ions. Thus, charge could flow without the metal ions mixing.

Q6.

a) Write a half equation for the reaction that occurs at the cathode in the Daniell cell.
b) Use the data in Table 14.2 to explain why a cell with an aqueous electrolyte can never generate more than 2.07 V.

14.6 THE NERNST EQUATION

As electrochemical cells rarely work under exactly standard conditions, the Nernst equation (named after Walther H. Nernst, 1864–1941) allows us to see how ion concentration and temperature variations affect the voltage generated by the cell.

The Nernst equation can be summarized as follows:

$$E = E^{\theta} - \frac{RT}{nF} \ln Q \tag{14.1}$$

and for a reaction at equilibrium:

$$\mathrm{aA + bB \rightleftharpoons cC + dD,} \tag{14.2}$$

where:

$$Q = \frac{[C]^c [D]^d}{[A]^a [B]^b} \tag{14.3}$$

Q7. The overall reaction for the Daniell cell can be shown as follows in eqn 14.4:

$$\mathrm{Zn_{(s)} + Cu^{2+}{}_{(aq)} \rightleftharpoons Zn^{2+}{}_{(aq)} + Cu_{(s)}} \tag{14.4}$$

a) Define "standard conditions" for electrochemical cells.
b) Write an expression for Q for the reaction in the Daniell cell.
c) Using the Nernst equation, show that, under standard conditions, $E = E^{\theta}$.
d) The Daniell cell is set up under standard conditions, **except** that the copper sulfate solution is diluted to 0.10 M. Calculate the E for the cell under these conditions.
e) Suggest two changes that you could make to a Daniell cell to **increase** the voltage output and explain your choices.

Using the Nernst equation, we might be convinced that if the value for $Q = 1$, then a temperature change will have no effect on the E of the cell. However, we need to think about how an increase in temperature will affect the position of this equilibrium, and hence the value for Q.

$$\Delta G = -nFE \tag{14.5}$$

As the electrode potential for the reaction is positive, the reaction is feasible (*i.e.*, it has a negative ΔG).

But, $\Delta G = \Delta H - T\Delta S$.

Q8.

a) Why would you expect ΔS to be small for the reaction in the Daniell cell?

So, we can say $\Delta G \approx \Delta H$.

b) How will an increase in temperature affect the position of equilibrium for the Daniell cell reaction?
c) Predict what will happen to the voltage of the Daniell cell as the temperature increases.

14.7 ZINC ALKALINE ELECTROCHEMICAL CELLS

In more recent times the alkaline cell has used zinc as the anode and manganese IV oxide as the cathode. The electrolyte is aqueous potassium hydroxide. Such cells offer a higher energy density than Daniell cells.

$$ZnO_{(s)} + H_2O_{(l)} + 2e^- \rightarrow Zn + 2OH^-_{(aq)} \quad E^\theta = -1.28 \text{ eV} \tag{14.6}$$

$$2MnO_{2(s)} + H_2O_{(l)} + 2e^- \rightarrow Mn_2O_{3(s)} + 2OH^-_{(aq)} \quad E^\theta = +0.15 \text{ eV} \tag{14.7}$$

Q9. Use the standard electrode potential values in eqn 14.6 and 14.7 to calculate the voltage generated when the two half cells are connected.

14.8 ZINC: GALVANIZING AND THE SACRIFICIAL ANODE

The majority of zinc extracted these days is used to galvanize iron and steel. The zinc coating protects the iron in two ways: firstly, by preventing water and oxygen from getting to the iron, thus protecting it from oxidation. Zinc itself is oxidized but the layer of zinc oxide effectively seals off the rest of the zinc and prevents further corrosion. Secondly, if the zinc coating is chipped and the iron is exposed, the greater reduction potential of the zinc electrode prevents the iron from oxidizing by behaving as a sacrificial anode.

Q10. Refer back to Table 14.2. Using appropriate half equations, show how the zinc coating in contact with iron prevents further oxidation of the exposed iron.

14.9 SOME ZINC BIOCHEMISTRY

Zinc's role in biochemical systems is, of course, determined by its chemical reactivity. Firstly, it has only one stable oxidation state and, secondly, it is readily soluble in aqueous solutions (unlike iron, it is not overly sensitive to the amount of dissolved oxygen). Recent research has revealed that zinc is vital in enzymes called transcription factors. It appears that the Zn^{2+} ion forms co-ordinate bonds with two histidine and two cysteine amino acids in these enzymes that are responsible for DNA transcription. The geometry of the zinc amino acid complex ensures that the enzyme has the correct shape to bond with a specific section of the DNA helix. Given this key role, it is not surprising that a zinc deficiency in the diet can have a serious impact on normal development.

Q11.

a) Suggest why the role played by zinc in enzymes is largely structural rather than as a catalytic centre, such as copper or iron might form.
b) *E. coli* grown on a special growth medium has been shown to concentrate zinc ions by a factor of 2000 times. Given that the internal concentration of zinc ions is 0.2 mM and that the maximum volume of a single *E. coli* cell is 1.8×10^{-15} dm^3, calculate the maximum number of Zn^{2+} ions per cell.

14.10 CONCLUSIONS

Living organisms seem to work particularly hard at securing certain elements from their surroundings and few are selected for more vigorously than zinc. Biologically, it seems to have developed a strong regulatory role. It has been shown to be present in a vast number of enzymes, including oxidoreductases, transferases, hydrolyases, ligases and transcription factors. In industry zinc plays a major role in electrochemical cells and the galvanizing of steel amongst others. Zinc may be rather less flamboyant than its periodic neighbor, copper, with regards to its colour and redox chemistry but, with the exception of iron, no other 'd' block element is required in more abundance than zinc in living organisms.

REFERENCES

1. P. J. White and M. R. Broadley, Biofortification of crops with seven mineral elements often lacking in human diets – iron, zinc, copper, calcium, magnesium, selenium and iodine, *New Phytologist*, 2009, **182**(1), 49–84.
2. D. J. Weisl, T. F. D. Mason, F. J. Zhao, G. J. D. Kirk, B. J. Coles and M. S. A. Horstwoord, *New Phytologist*, 2005, **165**(3), 703–710.

Answers to question in this chapter are on pages 214–216.

CHAPTER 15

Mercury

Sometime around the 4th century CE the Chinese alchemist Ko Hung had this to say about mercury:[1]

> *"After grass and wood have been burnt, they become ashes, but cinnabar can be changed into mercury by heating over a fire, and vice versa. It is far different from the ordinary vegetable substance so it can make people immortal."*

Mercury has always provoked a sense of wonder and so it is not surprising that it was to play a central role in alchemy. It is said that the mausoleum of the Chinese emperor Qin Shi Huang (259–210 BCE) contains rivers of mercury representing the rivers of his realm; although, even today, the site has yet to be excavated. It is even possible that he may have drunk mercury in an attempt to achieve the immortality mentioned by Ko Hung 600 years later. Not surprisingly he lived to the ripe old age of 49.

In medieval Europe mercury was known as quicksilver (*quick* in this case implying alive in some sense) and, with sulfur and salt, was again considered by Paracelsus to be one of the three fundamental substances.[†] Given such mystical beginnings, it is ironic that mercury was to play a central role in some of the most important scientific

[†]Paracelsus had a strong belief in the Christian idea of the Holy Trinity. For him, it was important to add a third element to mercury and sulfur so that the material world was analogous to the spiritual. In Chapter 5 on calcium you will see that even the great Issac Newton wasn't immune to such "adjustments".

Around the World in 18 Elements
By David A. Scott

Published by the Royal Society of Chemistry, www.rsc.org

developments. It was an important component in Michael Faraday's first electro-magnetic induction motor—the liquid metal enabling an electrical circuit to be complete, whilst allowing the wire to move. Mercury II oxide was the source of Joseph Priestley's "dephlogisticated air", later re-conceptualized as oxygen, which was to be pivotal in restructuring chemistry as a modern science. It was also important in the developing technology of instrumentation, including barometers and thermometers, and is still used to quantify the unit of electrical resistance, the ohm.

Today, students of chemistry are very unlikely to see liquid mercury in the lab as it is highly toxic, but internet video clips abound showing iron bolts and blocks of lead floating on mercury and it continues to fascinate.

15.1 LIQUID SILVER

The symbol for mercury, Hg, comes from the Latin *Hydrargyrum*, meaning "liquid silver", which is a pretty good description of its appearance, if not its chemistry. You can find it at the bottom right of the 'd' block in the periodic table, beneath cadmium. Although all three members of group 12 have relatively low melting and boiling points, mercury is unique in being a liquid at room temperature. At 25 °C, the metal has a vapour pressure of 2.5×10^{-1} N m^{-2}.

Q1.

a) Explain what is meant by the term *vapour pressure*.
b) What would happen to a sample of mercury if it was placed in a sealed iron container and the air evacuated to a pressure below 2.5×10^{-1} N m^{-2}.

As with all group 12 metals, the preferred oxidation state is M^{2+} but mercury is alone in forming the $Hg_2{}^{2+}$ ion. It seems that even though the bonding in liquid mercury is metallic, *i.e.*, delocalized electrons, in the $Hg_2{}^{2+}$ species, two Hg atoms are covalently linked with a bond length of about 0.3 nm. The involvement of the $6s^2$ electrons in a dative covalent bond helps explain mercury's poor electrical conductivity in comparison with other metals.

15.2 MERCURY AND ALCHEMY

Mercury (Figure 15.1) was a very special substance to alchemists in China, the Middle East and medieval Europe. For the Chinese

☿

Figure 15.1 Alchemical symbol for mercury.

alchemists, it was associated with Yin, the Moon and the feminine principle to be contrasted with Yang, the Sun and the masculine principle represented by sulfur. The Chinese believed that understanding alchemy would reveal the secret of immortality. Europeans were more interested in the belief that alchemy would enable the synthesis of gold from baser metals. Needless to say, nobody lived for ever in China, or got rich turning lead into gold in Europe. However, mercury and its compounds were to play a major part in the development of modern science and technology!

15.3 CINNABAR

Mercury is comparable in abundance in the Earth's crust to silver and about 50 times more abundant than gold. Its moderate reactivity means that it is rarely found in its elemental form but there are stories of it flowing from fissures in rocks during volcanic activity, the heat of the lava decomposing cinnabar into its constituent elements. What is certainly true is that its volatility means that it is present in the air in tiny but measurable quantities. The majority of metallic mercury is extracted from cinnabar (largely mercury II sulfide) by heating in air (eqn 15.1).

$$HgS_{(s)} + O_2 \rightarrow Hg_{(l)} + SO_{2(g)} \quad (15.1)$$

Q2. If 78% by mass of a sample of cinnabar was shown to be pure mercury, what percentage of the cinnabar sample is mercury II sulfide?

15.4 MERCURY II OXIDE AND THE BIRTH OF MODERN CHEMISTRY

Joseph Priestley was to make a number of contributions to 18th century science and politics but, as a chemist, it is his discovery of "dephlogisticated air" that was to prove most significant. We have already seen that mercury had a particular fascination for the

Table 15.1 $H^\theta{}_f$ and S^θ values for the decomposition of mercury II oxide, forming mercury.

Substance	$\Delta H^\theta{}_f$/kJ mol^{-1}	S^θ/J mol^{-1} K^{-1}
Hg	0.0	76.0
O_2	0.0	102.5
HgO	−90.8	70.3

alchemists and, in the later part of the 18th century in Europe, it was the alchemists who were most involved in ideas and experiments concerning the transformation of matter. It had been known for some time that mercury II oxide would decompose to produce mercury at moderately high temperatures but it seems that Priestley was the first to collect the gas given off when the oxide decomposed. His experiments using this gas are well known: it relighted a glowing splint and allowed a mouse to stay alive for nearly three times as long as another mouse kept in the same volume of air. It is worth looking at mercury's reactivity with oxygen as it illustrates some interesting aspects of chemical thermodynamics and kinetics. Consider the reaction Priestley used to isolate oxygen (eqn 15.2):

$$2HgO_{(s)} \rightleftharpoons 2Hg_{(l)} + O_{2(g)} \tag{15.2}$$

Q3.

a) Using the data in Table 15.1 and the relationship: $\Delta G = \Delta H - T\Delta S$, calculate the temperature at which the decomposition of mercury II oxide becomes feasible.
b) At temperatures of about 350 °C, after heating for a number of hours in a closed system, some of the mercury turns red. Explain this observation.
c) At room temperature, mercury does not react with oxygen. Comment on this observation in terms of both thermodynamics and kinetics.

It is ironic that mercury, which was such an important and almost mystical substance to the alchemists, was to play such a significant role in the overthrow of alchemy and the development of chemistry as we would now recognize it.

15.5 MERCURY AND THE CASTNER–KELLNER PROCESS

The electrolysis of brine is an important industrial process. Starting with one cheap and abundant raw material and producing no less than three important raw materials makes it something of a model for industrial processes. Practically speaking, there is one major problem to be overcome during the electrolytic process: the chlorine produced at the anode will react with sodium hydroxide to produce a sodium chlorate I solution. Thus, keeping the sodium hydroxide formed in the aqueous electrolyte away from the chlorine formed is the problem to be overcome. A number of engineering solutions have been developed to overcome this, one of which uses a flowing mercury cathode developed independently by Castner and Kellner who, rather than argue about patents, went into business together. Now, if you have completed GCSE chemistry, the chances are you have seen a video clip showing a flowing mercury cell. During the process, we are told that **sodium** forms at the cathode! The sodium/mercury amalgam flows out of the cell, where it reacts with water to produce sodium hydroxide and hydrogen gas, thus returning the mercury to the cell.

But hold on! What about the reactivity series! Why all of a sudden is sodium produced at the cathode and not hydrogen?! I confess that, when I fielded this question for the first time many years ago, I was stumped. You might not be surprised to find out that the "metal reactivity series" is something of an over-simplification. It is the thermodynamics of the formation of the sodium/mercury amalgam that is important. Amalgams are generally thought to be solutions of metals in mercury, but there is some evidence that the association of sodium with mercury has clear stoichiometric ratios: Na_5Hg_8 and Na_3Hg are two suggested "intermetallic compounds". The formation of these compounds is exothermic, and thus their formation is more thermodynamically feasible than the formation of hydrogen gas at the surface of the mercury. In addition, the formation of these stoichiometric amalgams means that the sodium cannot react with the aqueous solution, and thus the sodium is removed into another part of the process, where it reacts with water and a graphite catalyst to produce sodium hydroxide and hydrogen gas. When graphite electrodes are used, no such amalgam can form, and thus the thermodynamics favours the production of hydrogen at the cathode.

Q4. Write down equations for the following reactions:

a) The oxidation of chloride ions at the anode.
b) The formation of sodium amalgam at the cathode.
c) The reaction of the amalgam with water.

15.6 THE MERCURY CELL

Mercury has also been used in voltaic cells.

Q5.

a) Use the data in Table 15.2 to write an equation for the overall reaction between the two half cells, and hence calculate the potential difference generated by a single cell under standard conditions.
b) Using the relationship: $\Delta G = -nFE^{\theta}$, calculate the free energy change for this reaction.
c) Using the data in Table 15.3, calculate the standard enthalpy of reaction for the reaction in part a).
d) Compare and comment on your answers to parts b) and c). Why might you expect the values to be similar?

15.7 MERCURY: POISON OR CURE?

Mercury and its compounds range in toxicity depending on the oxidation state of mercury and whether or not it is in the form of an organometallic compound. "Corrosive sublimate" was made by heating mercury II sulfate and sodium chloride. The mercury II chloride produced is extremely toxic, and perhaps by trial and error, it was found that if the mercury II chloride was mixed with more mercury to form mercury I chloride (Hg_2Cl_2), also known as calomel, a less toxic substance resulted. Mercury I chloride was used as a laxative and also to treat syphilis, which, although endemic in Europe, is thought by some to have become more of a problem after the Europeans returned from the New World with a more potent version

Table 15.2 Standard electrode potentials for the mercury cell.

Half cell	*Standard electrode potential/eV*
$ZnO + H_2O + 2e \rightarrow Zn + 2OH^-$	−1.260
$HgO + H_2O + 2e \rightarrow Hg + 2OH^-$	+0.098

Table 15.3 H^{θ}_{f} for the mercury cell.

Substance	*ΔH^{θ}_{f}/kJ mol^{-1}*
HgO	−90.8
ZnO	−348.3

Table 15.4 Standard electrode potentials for mercury salts.

Half cell	*Standard electrode potential/eV*
$Hg_2{}^{2+} + 2e \rightleftharpoons 2Hg_{(l)}$	+0.789
$2Hg^{2+} + 2e \rightleftharpoons Hg_2{}^{2+}$	+0.920

of the bacteria. Treatment with these compounds must have been a question of which had the stronger constitution: the bacterial organism responsible for the infection or the patient.

Evidence for the relative stability of mercury salts can be obtained from standard electrode potentials, as shown in Table 15.4.

Q6.

a) Write an equation for the disproportionation of $Hg_2{}^{2+}$ ions and calculate an E^θ cell value for the process. What does this value tell you about the feasibility of disproportionation in this case?
b) Using the relationship: $nFE^\theta = RT\ln K$, calculate the ratio of $[Hg^{2+}]$: $[Hg_2{}^{2+}]$ in aqueous solution at equilibrium and 298 K.

There may be a clue to the biochemical mechanism by which mercury rather indiscriminately does its worst in the fact that mercury tends to accumulate in hair. In Chapter 4 on sulfur we noted that the human hair protein has a particularly high sulfur content and it is known that mercury II ions have an extremely high binding affinity for sulfur. (The solubility product for mercury II sulfide is 2×10^{-53} mol dm^{-3}!) Thus, any mercury II ions will bind irreversibly with sulfur-containing amino acids and disrupt protein function. As just about all proteins have some sulfur-containing amino acids, one might expect a considerable level of disruption. In the evolutionary arms race, however, bacteria are the masters and the organism *Pseudomonas aeruginosa* has evolved an enzyme that is capable of reducing mercury II ions to mercury atoms. This raises the possibility of detoxifying areas with high mercury II concentrations using cultures of these organisms.

15.8 ORGANOMERCURY

Organomercury compounds constitute mercury atoms covalently bonded to carbon. They are considered to be even more destructive than inorganic mercury salts because the hydrophobic nature of the organic groups means that they are far more able to cross cell membranes and, most seriously of all, the blood–brain barrier. The most infamous case of organic mercury poisoning took place in Japan in 1956. In the western coastal village of Minamata a mysterious illness started affecting people who were experiencing psychotic episodes before falling into a coma. It turned out that the dumping of mercury waste into the local sea found its way into the food chain. Perhaps up to 900 people died from mercury poisoning and a further 2250 were badly affected.

15.9 CONCLUSIONS

The usage of mercury is, understandably, under stricter regulations than it used to be. However, in some parts of the world it is still used in the extraction of gold as it effectively dissolves the gold to form an amalgam, from which it can be removed. In general mercury is now very carefully monitored. A concentration of 0.2 $\mu g\ g^{-1}$ of whole blood is thought to be enough to produce neurological symptoms. A mercurial personality, appropriately enough, is one that is erratic, unpredictable and even a little unstable. It is not difficult to see where these adjectives came from.

REFERENCE

1. S. F. Mason, *A History of the Sciences*, 1962, Collier Books, New York, p. 77.

Answers to questions in this chapter are on pages 217–218.

CHAPTER 16

Manganese

Element number 25 sits between chromium and iron in the periodic table and is, perhaps surprisingly, the twelfth most abundant element in the Earth's crust, frequently occurring along with iron in minerals. As the metal, manganese is used extensively for its alloying properties and there are few steels that don't include at least some manganese. It also has an extremely varied chemistry and is able to adopt oxidation states of $+2, +3, +4$, $+6$ and $+7$. In the chemistry lab manganate VII compounds are regularly used in quantitative analysis due to a) their strength as oxidizing agents and b) their clear changes in colour that accompany changes in oxidation number. What is less commonly known is that manganese chemistry is responsible for the production of oxygen from water during the process of photosynthesis; we have manganese to thank for the air we breathe! It is also possible that manganese's ability to adopt various oxidation states enables it to be involved in geochemical cycles and, for the same reason, it has played a significant role in the development of electrochemical cells. The existence of manganese nodules at the bottom of the sea and other inland stretches of water has raised the possibility of mining the resource in the future.

16.1 MANGANESE AND REDOX CHEMISTRY

Manganese exhibits all of the typical properties of a transition metal. It is stable in a number of oxidation states, forms coloured complexes

Around the World in 18 Elements
By David A. Scott

Published by the Royal Society of Chemistry, www.rsc.org

and it catalyzes reactions. Perhaps the first example of a catalytic reaction students are likely to see is the decomposition of hydrogen peroxide using a manganese IV oxide catalyst. The reaction is summarized in eqn 16.1:

$$2H_2O_{2(aq)} \xrightarrow{MnO_{2(s)}} 2H_2O_{(l)} + O_{2(g)} \tag{16.1}$$

It might be surprising to find out that the mechanism for this reaction is still not completely understood. It is thought that an initial step involves manganese dioxide in contact with hydrogen peroxide being reduced to Mn^{2+} and Mn^{3+} states, possibly to generate a superoxide species before being oxidized back to $+4$ oxidation state. This is an example of a heterogeneous catalyst, which is in a different phase to the reactants. Chemists like heterogeneous catalysts as they tend to be easier to separate from the reaction mixture at the end of the reaction. On the negative side, mechanisms involving heterogeneous catalysts tend to be less well understood.

Q1. Explain the general differences between the different states of matter and different phases.

16.1.1 Redox Titrations Involving Manganate VII

Manganese is the only transition metal in period 4 that can exist in a $+7$ oxidation state. This makes for one of the most potent oxidizing agents that can be used easily in aqueous solution.

Q2.

a) A solution of $KMnO_4$ is made by dissolving 1.58 g of the solid in a solution of 500 cm^3. Calculate the concentration of this solution in mol dm^{-3}.
b) Combine the two half equations shown in eqn 16.2 and 16.3 to give a stoichiometric equation for the complete oxidation of oxalic acid and calculate the standard electrode potential for the reaction.

$$MnO_4{}^-{}_{(aq)} + 8H^+{}_{(aq)} + 5e^- \rightarrow Mn^{2+}{}_{(aq)} + 4H_2O_{(l)} \quad E^{\theta} = +1.51 \text{ eV} \tag{16.2}$$

$$2CO_{2(aq)} + 2H^+{}_{(aq)} + 2e^- \rightarrow H_2C_2O_{4(aq)} \quad E^{\theta} = +0.39 \text{ eV} \tag{16.3}$$

c) 25.00 cm^3 of oxalic acid solution required 15.25 cm^3 of the prepared potassium manganate VII solution to reach an end point in this reaction. Calculate the concentration of the oxalic acid solution to the appropriate number of significant figures.
d) Explain why there is no need for an indicator to judge the end point in this reaction.

An interesting twist to this reaction is that the reduced $Mn^{2+}{}_{(aq)}$ ions produced in the reaction actually catalyze the forward reaction. In the laboratory a temperature between 50–60 °C is required to start the reaction but, once initiated, the reaction speeds up before slowing down again. One way to investigate the extent of this self-catalyzing reaction would be to chelate the Mn^{2+} ions as they form using, for example, EDTA.

e) Explain what is meant by a chelating agent.
f) What do you think a rate of reaction profile of rate *vs.* time might look like for a self-catalyzing reaction of this type?

16.2 MANGANESE AND BIOGEOCHEMICAL CYCLES

So much for manganese chemistry in the lab! Given its relative abundance in the Earth's crust, it is not surprising that it is involved in environmental redox chemistry. In the analysis of a water sample from Westbere Lake near the city of Canterbury, Kent, on 15th October 2013 a value of 60.7 μg dm^{-3} manganese ions was recorded, which is nearly three times the concentration of aluminium ions in the same sample.†

In the Black Sea, soluble Mn^{2+} ions are oxidized by atmospheric oxygen at the sea's surface:

$$2Mn^{2+}{}_{(aq)} + O_{2(aq)} + 2H_2O_{(l)} \rightarrow 2MnO_2 + 4H^+{}_{(aq)} \quad (16.4)$$

The manganese IV oxide produced is insoluble and so sinks down to lower levels, where it can oxidize hydrogen sulfide ions to sulfur (eqn 16.5):

$$HS^-{}_{(aq)} + MnO_2 + 3H^+{}_{(aq)} \rightarrow S_{(s)} + Mn^{2+}{}_{(aq)} + 2H_2O_{(l)} \quad (16.5)$$

In doing so, it regenerates Mn^{2+} ions that can start the cycle again. This process is almost certainly essential for the aquatic habitat in the upper parts of the sea as high concentrations of hydrogen sulfide are

†Water analysis performed by South East Water Scientific Services, Frimley Green, Surrey, on 15/10/2013, with thanks to Richard Brown, laboratory manager.

Table 16.1 Standard electrode potentials for $MnO_2 + 4H^+ + 2e \rightleftharpoons Mn^{2+} + 2H_2O$, $NO_3^- + 2H^+ + 2e \rightleftharpoons NO_2^- + H_2O$ and $NO_3^- + 10H^+ + 8e \rightleftharpoons NH_4^+ + 3H_2O$.

Half equation	*Standard electrode potential/eV*
$MnO_2 + 4H^+ + 2e \rightleftharpoons Mn^{2+} + 2H_2O$	
$NO_3^- + 2H^+ + 2e \rightleftharpoons NO_2^- + H_2O$	+0.94
$NO_3^- + 10H^+ + 8e \rightleftharpoons NH_4^+ + 3H_2O$	+0.87

highly toxic to many organisms. The overall result, however, is an oxygen concentration that drops off extremely rapidly in waters such as the Black Sea, which is thought to be anoxic at depths below 100 m. It is not only hydrogen sulfide that is influences by Mn^{2+} ions.

Q3.

a) Using the electrode potential data in Table 16.1, discuss the impact of the Mn^{2+}/MnO_2 equilibrium on the availability of soluble nitrogen species in an aquatic environment. You should write half equations to help illustrate your reasoning.
b) How predictive are standard electrode potential calculations in marine environments? Explain your answer.

Hopefully, Q3 illustrates that redox chemistry doesn't just happen in the lab. It is central to understanding biogeochemical cycles!

16.3 SOME MORE REDOX REACTIONS OF MANGANESE IONS IN SOLUTION

Manganese II salts are slightly soluble in water and form the $[Mn(H_2O)_6]^{2+}$ ion. This ion is relatively stable to redox reactions and, in many ways, mimics the weakly complexing group II metal ion calcium. Indeed, the range of the manganese concentration in bones can be anything from 0.2 to 100 ppm. It could be that excess manganese II is stored in bone tissue because the body can metabolize it, like calcium, if needs be. The hexaaquamanganese II ion reacts with alkali to form a beige-coloured precipitate, which slowly turns dark brown on exposure to air.

Q4.

a) Give the s,p,d electronic configurations of Mn and Mn^{2+} and suggest why you might expect the Mn^{2+} ion to be reasonably stable.

b) Write a balanced equation for the reaction of the hexa-aquamanganese II ion with alkali.
c) Suggest the reason for the darkening of the beige precipitate on exposure to air.

16.4 MEASURING DISSOLVED OXYGEN: THE WINKLER TITRATION

Dissolved oxygen in water is a key factor in determining biodiversity. By definition, all aerobic organisms require oxygen. The oxygen present in water can come through a process of diffusion from the air or as a waste product from photosynthesis. The Winkler titration is a method of measuring the amount of dissolved oxygen and takes advantage of manganese's variable oxidation states. The procedure is outlined below.

An excess amount of $MnSO_4$ is added to the measured volume of the water to be tested followed by an alkaline solution of potassium iodide (also in excess). The Mn^{2+} ions form a precipitate with the OH^- ions whereupon the manganese II hydroxide is oxidized by the dissolved oxygen to form a precipitate of manganese III hydroxide, which is stable in alkali. Once the manganese III hydroxide precipitate has settled, it is dissolved in just enough sulfuric acid to release the Mn^{3+} ions into solution whereupon the reaction shown in eqn 16.6 occurs:

$$2Mn^{3+}{}_{(aq)} + 2I^-{}_{(aq)} \rightarrow 2Mn^{2+}{}_{(aq)} + I_{2(aq)} \quad (16.6)$$

The amount of iodine produced can now be quantified by a thiosulfate titration. As the amount of I_2 produced is exactly half the amount of Mn^{3+} reduced, we can calculate the quantity of oxygen dissolved in the sample.

Q5.

a) Write balanced equations for the following reactions described in the above procedure:
   (i) The precipitation of manganese II hydroxide in alkaline solution.
   (ii) The oxidation of manganese II hydroxide to manganese III hydroxide by dissolved oxygen.
   (iii) The reaction between iodine and thiosulfate ions.
b) A 100 cm^3 sample of river water was tested following the procedure outlined above. 8.55 cm^3 of a standardized

0.100 M sodium thiosulfate was required to reach an end point. Calculate the concentration of oxygen dissolved in the water in mg dm^{-3}.

c) A healthy river is considered to be one that has a dissolved oxygen concentration of between 6 and 10 mg dm^{-3}. Does the sample in part b) come from a healthy river?

d) Is the end point of 8.55 cm^3 a suitable volume on which to base a titration? Explain your answer.

e) Suggest a reason why many chemical industries that require large volumes of water may prefer water with a low dissolved oxygen content. How might this be achieved?

16.5 SOME MORE USEFUL MANGANESE COMPOUNDS

As already noted, the main usage for manganese is in steel alloys, where it both contributes to the hardness of the alloys and is an effective scavenger of impurities. For example, manganese will react with both sulfur and oxygen impurities to form sulfides and oxides, respectively. We have seen in the chapter on zinc (Chapter 14) that manganese IV oxide acts as the cathode in dry electrochemical cells. Manganese IV oxide is also used in the preparation of chlorine gas, as can be seen in the following reaction:

$$MnO_{2(s)} + 4HCl_{(aq)} \rightarrow MnCl_{2(aq)} + Cl_{2(g)} + 2H_2O_{(l)} \quad (16.7)$$

Q6.

a) Identify the changes in oxidation numbers in eqn 16.7.

b) What volume of $Cl_{2(g)}$ measured at room temperature and pressure would be produced if excess concentrated hydrochloric acid was added to 450 kg of $MnO_{2(s)}$. Note that 1 mole of gas occupies 24 dm^3 at room temperature and pressure.

Potassium manganate VII (which you will still see referred to as potassium permanganate) is a widely used chemical in industry, both as an oxidizing agent, an analyte and, occasionally, as a treatment for water supplies. It is produced from manganese IV oxide in a two-step process (eqn 16.8 and 16.9).

1. $2MnO_{2(s)} + 2KOH_{(aq)} + O_2 \rightarrow K_2MnO_4 + 2H_2O$ (16.8)

2. $2K_2MnO_4 + 2H_2O \rightarrow 2KMnO_4 + 2KOH + H_2$ (16.9)

The second step of this reaction is carried out by electrolysis. This is done to regenerate the potassium hydroxide solution and produces hydrogen at the cathode. This may seem like an energy-intensive way to carry out a reaction that can occur spontaneously on the addition of acid (eqn 16.10).

$$3MnO_4^{2-} + 4H^+ \rightarrow 2MnO_4^- + MnO_2 + 2H_2O \qquad (16.10)$$

Q7.

a) Identify oxidation numbers +4, +6 and +7 in the reactions shown in eqn 16.8–16.10.
b) Write anode and cathode reactions for the electrolysis of K_2MnO_4 in alkaline solution.
c) Suggest why the more energy-intensive process of electrolysis is used to produce $KMnO_4$ rather than reacting with acid.

16.6 MANGANESE AND THE EVOLUTION OF PHOTOSYNTHESIS

Geological records show that about 2.8×10^9 years ago the Earth's atmosphere started to change due to a new biochemical process. It is generally accepted that cyanobacteria started to split water. (See Figure 17.1 in Chapter 17.)

$$2H_2O \rightarrow 4H^+ + O_2 + 4e^- \qquad (16.11)$$

The protons generated were to provide the necessary proton gradient for the generation of ATP in mitochondria and the oxygen was a waste product. This waste product, however, would enable the evolution of aerobic respiration, which is a far more efficient respiratory process, producing something like eighteen times as much energy per unit mass of fuel in comparison to anaerobic respiration. It is only through the evolution of aerobic respiration that multicellular organisms were able to develop. It's hard to think of a more pivotal evolutionary piece of biochemistry!

In 1937 it was shown that low photosynthetic activity in algae deficient in manganese could be restored by the addition of manganese II salts in less than an hour. Subsequent studies, though still ongoing, have shown manganese to play a central role in the activity of photosystem II, the light-dependent part of photosynthesis.

16.7 CONCLUSIONS

In a sense we have come full circle: we started this chapter by recalling a simple laboratory preparation of oxygen from the manganese IV oxide catalyzed decomposition of hydrogen peroxide. It now seems that evolution stumbled on the importance of manganese as a catalyst for producing oxygen some 2.8 billion years before us. Many scientists believe that the cosmos may be teaming with life equivalent to bacterial-like organisms but that the evolution of multicellular life might be a vanishingly small event in cosmic biochemistry. Among a whole range of requirements for the development of complex life, we may also need to add "*a ready supply of manganese*" to the list.

Answers to questions in this chapter are on pages 219–221.

CHAPTER 17

Oxygen

Oxygen: the life giver, the air we breathe, the element whose discovery reshaped the whole of chemistry.

Any alien life-form analyzing the Earth's atmosphere from a distance would instantly notice something strange: it is a system in a steady state but far from thermodynamic equilibrium. We have already seen that defining life is not easy but all life that we know of has one thing in common: work is done to keep the living thing from reaching complete equilibrium with its surroundings. To reach this equilibrium would mean that there is no available free energy to address the inevitable increase in entropy, or to put it more bluntly: death. Oxygen's role in this whole process is an interesting one. Ironically, the "life giver" was responsible for the Earth's first mass pollution event. The appearance of dioxygen in the atmosphere was to spell the end for obligate anaerobes. But in doing so, evolution stumbled on an entirely new type of respiration: aerobic respiration. This was to enable far greater quantities of free energy to be made available. From that emerged multicellular life-forms, tissue specialization and a diversification of life, much of which is still uncatalogued today. And yet even aerobes have a complex relationship with oxygen. Being able to get access to more free energy comes at a price; aerobes make something of a Faustian pact with oxygen. It is a reactive element and certain forms of the element, called reactive oxygen species (ROSs), damage the very information-containing molecules that helped make us in the first place. The living systems' response to this has been to evolve a great number of mechanisms to

Around the World in 18 Elements
By David A. Scott

Published by the Royal Society of Chemistry, www.rsc.org

counteract this damaging effect, but they only delay the inevitable. All aerobes eventually equilibrate with their surroundings.

17.1 OXYGEN ABOVE AND BELOW

It's fairly common knowledge that oxygen (or more correctly, dioxygen) makes up about $\frac{1}{5}$ of the Earth's atmosphere or, more precisely, 20.8%. Because many younger students are so bound to the idea of oxygen as a gas, it is fun to see the looks of disbelief when I tell them that it also makes up about 47% of the Earth's crust, not as a gas, of course, but in chemical compounds. In fact, the real surprise is that there is any oxygen in the air at all and without photosynthesis, of course, there wouldn't be.

As any gardener will tell you, soil is not uniform stuff; it varies greatly with local geology and biology but, for the purposes of simplicity, let's consider the chemical composition of a "typical" sample of soil: removing organic material, water and dissolved gases, we are left with about 45% by mass of sand, silt and clay. This remaining dry mass in our sample has a composition shown in Table 17.1.

Q1.

a) Using the data in Table 17.1 calculate the total mass of oxygen in 10 kg of dry soil.
b) What colour might you expect the soil sample to be? Explain your answer.

The elemental composition of lunar soil is not dissimilar to our own, with oxygen and silicon comprising more than half of the mass, but it has disproportionately more iron than aluminium. This, along with other evidence, supports the theory that the moon was once part of a proto-Earth planet, which was split a part by a massive interplanetary collision over 4.6×10^9 years ago.

Table 17.1 The percentage by mass of the compounds and minerals found in a typical soil sample.

Compound/mineral	*Percentage by mass*
SiO_2	65
$CaCO_3$	12
$KAlSi_3O_8$	11
Fe_2O_3	7
TiO_2	3
MgO	2

17.2 OXYGEN: THE FIRST GLOBAL POLLUTANT

Figure 17.1 represents what is thought to be the variation of oxygen in the Earth's atmosphere over time. The units on the y-axis show the partial pressure of oxygen and can be roughly translated in to a percentage composition by volume of the air.

If life did evolve on Earth as early as 3.85×10^9 years ago, then we can see that it plodded along steadily but rather unspectacularly for 1.4×10^9 years without oxygen. We have seen in Chapter 4 on sulfur that living organisms may have thrived in and around hydrothermal vents at the bottom of the oceans and derived their energy chemotrophically. However, around 2.45×10^{-9} years ago evolution was to stumble on what must be one of the most significant events in the Earth's history. The set of chemical reactions collectively referred to as photosynthesis that was to split water was to be a game changer.

$$2H_2O \rightarrow O_2 + 4H^+ + 4e^- \quad E^\theta = -1.229 \text{ eV} \tag{17.1}$$

Q2. Using the relationship: $\Delta G^\theta = -nFE^\theta$, calculate the energy required to carry out the reaction in eqn 17.1 and cxplain why the reaction is not feasible under standard conditions.

This thermodynamically "difficult" reaction was to provide the reducing power to convert carbon dioxide into sugars and, in doing so, convert a proportion of the solar energy into a kinetically stable source of chemical energy. Aerobic respiration evolved to make use of

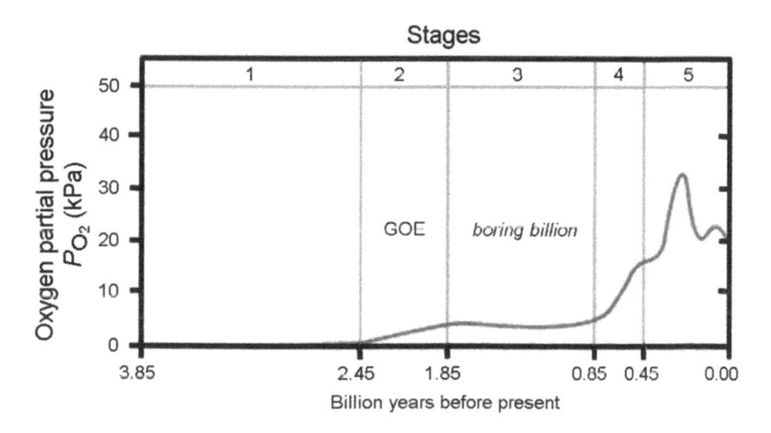

Figure 17.1 The variation of oxygen in the Earth's atmosphere over time (GOE = great oxidation event). Reproduced from Ref. 1.

this new source of chemical energy and changed the planet's surface for good.

$$CO_2 + 4H^+ + 4e^- \rightarrow (CH_2O) + H_2O \quad (17.2)$$

Q3. Explain why it is important for a viable source of chemical energy to be thermodynamically unstable but kinetically stable.

A particularly interesting trend on the graph in Figure 17.1 shows an oxygen peak corresponding to an oxygen percentage composition of over 30% at about 2×10^8 years ago. Paleontologists have found fossilized evidence for "giant" insects from this period. Insects and other arthropods rely on oxygen diffusing into their bodies, and so are naturally limited in size because surface-area-to-volume ratios decrease with size. But a 50% increase in the percentage oxygen in the air would allow a proportional increase in insect and arthropod size. A funnel web spider the size of a small cat could have been a real possibility!

17.3 THE "DISCOVERY" OF OXYGEN

The cast:

Joseph Priestley: Clergyman, chemist, political theorist, Yorkshireman.

Carl Wilhelm Scheele: Chemist, prolific discover of elements, especially given that he only lived to 44, Swedish.

Antoine-Laurent de Lavoisier: Nobleman, wealthy, chemist, French (loses his head in the final scene).

It is often said that there is something inevitable about scientific discovery and that, given the right conditions, certain advances will be made. Newton and Leibniz both independently came up with the calculus at about the same time. Martin Rees in his fourth Reith lecture in 2010 said about Einstein:

"Had he never existed, all of his insights would now have been revealed."

So it is, perhaps, not surprising that at least three scientists were experimenting with gases at about the same time at the end of the 18th century. For this reason, it is difficult to accredit an individual with the discovery of oxygen as all three scientists made a contribution. Priestley was to produce oxygen by decomposing mercury II oxide to the metal and oxygen, whereas Scheele was to get the same gas by thermally decomposing saltpeter ($NaNO_3$). However, it is often seeing the same thing in a different way that leads to what philosophers of science call a paradigm shift. Lavoisier met Priestley in Paris and they must certainly have talked about Priestley's "dephlogisticated air"; however, but it was Lavoisier who was to reconceptualize"dephlogisticated air" as the element oxygen. He was able to do this, in part, because his considerable personal wealth enabled the procurement of some of the best instrumentation of the day. He was able to show that when the element phosphorus burned in air it *gained* mass! Burning was previously seen to result in mass loss, as the rendering of logs of wood to a small pile of ashes clearly showed. Lavoisier was able to measure the increased mass of phosphorus with greater precision than had been previously accomplished and numerous repetitions confirmed that this increase in mass was always **proportional** to that mass of phosphorus he started with. Henceforth, chemistry was to become a truly quantitative science.

Q4.

a) Lavoisier measured out 15.60 g of phosphorus and placed it in a crucible and allowed it to react with the air. If the final mass of the product weighed 35.73 g, what was the mass increase?
b) Assuming that the added mass is due to oxygen, calculate the empirical formula of the oxide of phosphorus.
c) Lavoisier went on to investigate aluminium's reaction with air. If aluminium reacted with oxygen, what would be the final mass if 12.80 g of aluminium completely reacted with oxygen according to eqn 17.3:

$$4Al + 3O_2 \rightarrow 2Al_2O_3 \quad (17.3)$$

17.4 OXYGEN: "ACID FORMER"

Having discovered his new element, Lavoisier had to come up with a name (and, as it turned out, he didn't have much time)! He settled with a word derived from Greek that meant acid former and not without reason. It was well known that substances left in open air for a period of time became sour; wine, for example, and so he postulated that it was oxygen that was responsible for the phenomenon of this sourness or acidity. Some inorganic acids are shown in Table 17.2.

Q5.

a) What evidence in Table 17.2 supports Lavoisier's proposition that oxygen was responsible for the souring of substances left in open air.

Table 17.2 pK_a values of some inorganic acids.

Acid	*pK_a*
HNO_2	3.3
HNO_3	40
H_2SO_3	1.8
H_2SO_4	1000 +

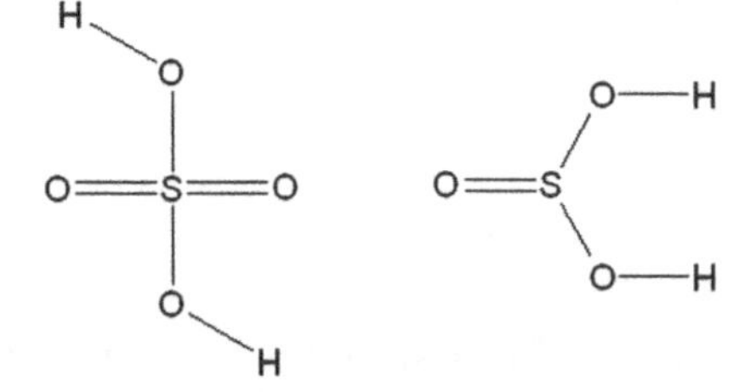

Figure 17.2 Structures of H_2SO_4 (left) and H_2SO_3 (right).

b) Sulfuric acid is considered to be a diprotic acid. Explain what this means and **sketch** an acid–base titration curve for the complete neutralization of H_2SO_4 by NaOH.
c) Calculate the pH of a 0.025 M solution of sulfuric acid stating any assumptions made.
d) Molecular structures of H_2SO_4 and H_2SO_3 are shown in Fig. 17.2. Give oxidation numbers for sulfur in each acid and, with reference to their structures, explain the differences in pK_a values.

17.5 OXYGEN AND THE OZONE LAYER

Most of the time, the use of the term *oxygen* is synonymous with molecular dioxygen (O_2). However, oxygen can also form the triatomic species O_3, called ozone. This is an even more potent oxidizing agent but tends not to form in the lower atmosphere. In the upper atmosphere, however, conditions are different and high energy ultraviolet (UV) radiation causes the O_2 molecule to split and the equilibrium in eqn 17.4 shifts to the right hand side forming ozone.

$$3O_{2(g)} \rightleftharpoons 2O_{3(g)} \ \Delta H^{\theta} = +285 \text{ kJ mol}^{-1} \quad (17.4)$$

The formation of this species is, to say the least, fortuitous, for developing life on the Earth's surface. Ozone absorbs the frequencies of UV radiation that no other molecular species in the atmosphere absorbs. By filtering out the more harmful frequencies of UV radiation centered on a wavelength of around 250 nm, ozone prevents it from reaching the Earth's surface, where it would break apart most organic molecules as soon as they formed. Life would simply not be able to get started. The first photosynthetic organisms almost certainly flourished at a depth in the oceans that protected them from damaging UV radiation. Once the oxygen had become concentrated enough to allow the ozone layer to form, life would have been able to exploit new habitats on land.

17.6 OXYGEN IN ORGANIC CHEMISTRY

A large proportion of carbon on Earth exists in a chemically reduced form, such as in the form of alkanes in crude oil. Carbon is also present in the atmosphere in its most oxidized form of carbon dioxide. But there are many important molecules in which carbon is in a partially oxidized state and a great number of these are essential to life. They include sugars, lipids, amino acids, hormones and others. Let's have a look at some examples (Figures 17.3 and 17.4).

Q6. Ribose is a sugar molecule that plays an important role in the structure of DNA.

a) Give the molecular formula of ribose and calculate the percentage mass of oxygen in the ribose molecule shown in Figure 17.3.
b) How many chiral centres are shown in the molecule?
c) What is the total possible number of optical isomers for this molecule?
d) Explain what is meant by the letter D in front of the name and how this letter is assigned.

The next structure is vitamin B_2 or riboflavin (Figure 17.4)

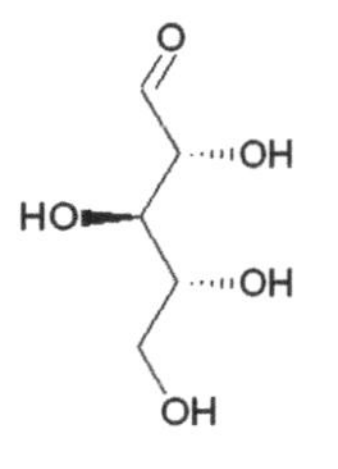

Figure 17.3 Structure of D-ribose.

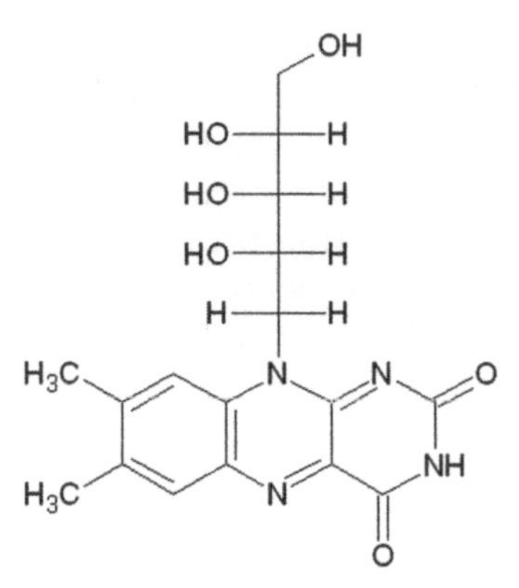

Figure 17.4 Structure of vitamin B_2, also known as riboflavin.

Q7.

a) Calculate the relative molecular mass of riboflavin and the percentage mass of oxygen present.
b) Which functional groups would you expect to be able to identify using IR spectroscopy?
c) A solution of riboflavin has a visible yellow colour. What structural feature of the molecule would you expect to account for this colour? Explain your answer.

17.7 REACTIVE OXYGEN SPECIES

It was noted in the introduction that, in adopting aerobic respiration, life has made something of a pact with oxygen. The greater energy available has allowed multicellular life-forms to evolve, leading to vastly more complex organisms and even culture. But along with oxygen comes ageing, and oxidative stress leads to molecular damage, a reduction in sperm quality for males and, eventually, death. Many of the products of aerobic reactions are reactive oxygen species (ROSs) that are even more potent than dioxygen and,amongst these, are the superoxide, peroxide and hydroxide radicals. In an attempt to counteract the damage that these radicals can do organisms have evolved a suite of enzymes that can quickly break down the most reactive species into less harmful ones. (One of these enzymes, superoxide dismutase, was considered in Chapter 9 on copper.) There also appears to be dietary measures that we can take to mitigate the effects of these radicals, which include naturally occurring molecules with antioxidant properties. One such molecule is vitamin C (Figure 17.5).

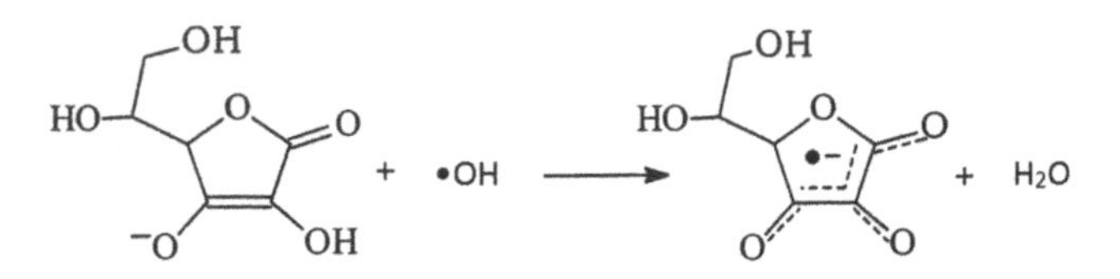

Figure 17.5 The ascorbate anion (from vitamin C, left) can react with free radicals to produce a less reactive ascorbate radical anion (right).

The ascorbate anion shown on the left of the equation in Figure 17.5 can react with a hydroxyl radical. In doing so it is able to delocalize the single electron around the five-membered ring, producing a less reactive ascorbate radical anion.

Q8.

a) The recommended daily allowance for vitamin C is between 70–90 mg per day. Calculate the number of moles this represents at the upper limit.
b) Identify the two chiral carbons in the vitamin C molecule.
c) Would you expect vitamin C to be fat soluble or water soluble? Explain your answer.

17.8 CONCLUSIONS

There's no getting away from it: multicellular life-forms need oxygen. Buried deep within the human genome resides a set of genes that carry out a process called glycolysis. This ancient relic allows even humans to get some energy anaerobically, but not for long as the build-up of lactic acid will soon testify. We are dependent upon oxygen.

REFERENCE

1. H. D. Holland, *Philos. Trans. R. Soc., B.*, 2006, **361**, 903.

Answers to questions in this chapter are on pages 222–223.

CHAPTER 18

Carbon

Carbon is present in the Earth's crust at about 480 ppm, putting it at number fifteen in the elemental abundance list, which doesn't seem particularly remarkable; however, it is carbon that forms more compounds than all of the other elements put together...far more! Even as the pure element, carbon appeals to the poetic (diamond) and the prosaic (graphite). But it is the simple and macromolecular compounds of carbon that have made such an astonishing impact on our planet. It is both the hardware and software of life. Biology is but a branch of carbon chemistry. To attempt to write a chapter about carbon chemistry that is anything more than cursory is to attempt the absurd. The number of possible carbon-based configurations is infinite and individuals may spend a lifetime working on the reaction mechanism of a single organic reaction.

The A level chemist preparing for the exams knows that both reagents and conditions are needed for certain organic synthetic procedures. This is not because chemistry examiners are a particularly sadistic lot, it is because something as seemingly insignificant as changing the composition of a solvent can change the products in a reaction. Carbon is such a versatile element that some reactions may involve a number of different mechanisms occurring simultaneously. This is why you will sometimes see a rate of reaction order of $\frac{1}{2}$ or $\frac{3}{2}$ with respect to a particular reactant. That said, analytical techniques, such as NMR and mass spectrometry, have now enabled structural details of highly complex molecules to be routinely determined and complex three dimensional carbon-based structures can now be

Around the World in 18 Elements
By David A. Scott

Published by the Royal Society of Chemistry, www.rsc.org

designed on the computer screen (*in silico*), synthesized in the lab (*in vitro*) and used in the body (*in vivo*).

For the **biochemist**, the early years of the 1950s were to produce some of the most remarkable breakthroughs. In 1953 Stanley Miller *et al.* at the University of Chicago, USA, showed that the amino acids alanine, glycine and aspartic acid, three of the building blocks of proteins, could be synthesized abiotically given the conditions that were thought to mimic those on early Earth. In the same year in Cambridge, UK, Crick and Watson *et al.* were to show that the hereditary principle was a chemical polymer: DNA. Sixty years after the discovery of both, genetic modification is a reality and discussions about the pros and cons of rice varieties that can be engineered to produce vitamin A are no longer the stuff of science fiction. At the same time, Rovers move over the surface of Mars looking for evidence of the sort of molecules—amino acids amongst them—that might provide evidence for life. Such is the infinite potential of the element carbon. Given this limitless fecundity, ironically perhaps, it is carbon in the form of carbon dioxide that is thought by many to be our most pressing environmental priority. In this final chapter I will look at some of these aspects of carbon.

18.1 CARBON: LIMITLESS POTENTIAL

Let's consider the versatility of carbon by looking at a small organic molecule of molecular mass = 88 g mol^{-1}: $\mathbf{C_4H_8O}$.

The elemental analysis of this compound will give you its empirical formula and, from there, the molar mass can be determined—these days by mass spectrometry—as a multiple of the empirical formula. However, the number of possible permutations involving these thirteen atoms means that, even once we have the molecular formula, we are still a long way from determining the structural formula. Consider the following:

Six molecules with the molecular formula C_4H_8O have been isolated and purified. They are all colourless liquids and are labelled A to F. A number of chemical "wet") tests are carried out on each and the results are summarized in Table 18.1.

Q1. Use the results in Table 18.1 to suggest structural formulae for molecules A to F that are consistent with both the molecular formula C_4H_8O and the chemical tests. Note that, in some cases, there will be more than one possible molecule. (You might need to brush up on a few of your functional group tests!)

Table 18.1 Chemical wet test carried out on molecules A to F.

	Test with 2,4-dinitrophenol	*Test with Tollens' reagent*	*Test with acidified potassium dichromate*	*Test with alkaline potassium manganate VII*	*Test with* PCl_5	*Iodoform test*
A	Orange precipitate	No reaction	No reaction	No reaction	No reaction	Yellow precipitate
B	Orange precipitate	Silver mirror	Orange to green	No reaction	No reaction	No reaction
C[a]	No reaction	No reaction	Orange to green	Purple to colourless	White fumes given off that turn moist blue litmus red	No reaction
D	No reaction	No reaction	No reaction	Purple to colourless	No reaction	No reaction
E	No reaction	No reaction	No reaction	No reaction	White fumes given off that turn moist blue litmus red	No reaction
F[b]	No reaction	No reaction	Orange to green	Purple to colourless	White fumes given off that turn moist blue litmus red	Yellow precipitate

[a]Has geometric isomers.
[b]Has optical isomers.

And that's just six! There is a freely accessible database of 34 000 compounds online, which has **nineteen** different compounds with this molecular formula (see: http://sdbs.riodb.aist.go.jp/sdbs/cgi-bin/direct_frame_top.cgi).

18.2 SPECTROSCOPIC ANALYSIS

The analytical organic chemist's job is not an easy one, as we can see, but there is no doubt that modern instrumental techniques that use the interaction of various parts of the electromagnetic spectrum with molecules have made life easier. IR and NMR spectroscopy have made structural interpretations much quicker, easier and also enabled the differentiation between isomers that are simply not possible with "wet" chemical tests, such as the ones used above.

Q2.

a) One of the molecules (A–F) was analyzed using IR and proton NMR spectroscopy. The main observations from each technique are summed up as follows:

- IR spectroscopy: showed a strong absorbance at about 1720 cm^{-1}. Apart from C–H bond vibrations, no other absorbances were recorded above 1500 cm^{-1}.
- Proton NMR analysis: showed three distinct proton peaks centred on 1, 2.1 and 2.4 ppm in the ratio of 3 : 3 : 2, respectively. The splitting patterns resolved into a triplet, singlet and a quartet, respectively.

On the basis of these observations identify which of the molecules A–F has been analyzed accounting for the main observations in each technique.

b) Compound **C** has two different **geometric isomers.** Draw and name both isomers.

You are probably getting the idea by now! The same database search for $C_8H_{14}O_2$ returns 41 hits and even this will only be a fraction of the possible structures with this formula.

Q3. How many different isomers do you think are possible with the molecular formula $C_{25}H_{52}$?

18.3 CARBON CHEMISTRY AND OPTICAL ISOMERISM

Carbon's ability to form four bonds in a tetrahedral arrangement gives rise to the last type of isomerism that A level chemists will encounter;

namely, optical isomerism. This is essentially the ability of carbon-based molecules to form non-superimposable mirror images of each other. If a carbon atom is bonded to four different groups, it can exist as two non-superimposable isomers, which we call optical isomers.

Q4. Compound F in Table 18.1 can exist as two optical isomers. Draw out a possible structure for compound F and identify the chiral carbon.

18.4 VITALISM AND ORGANIC CHEMISTRY

The Vitalist philosophy (which we first encountered in Chapter 2 on iron) had its origins in ancient Egypt and Greece, and believed that living organisms were qualitatively different from inorganic matter. Even as chemistry progressed to show that organic substances were composed of elements common with inorganic matter it was still believed that some other principle or essence was needed to bring life to organisms. The preparation of alcohol (ethanol) in some shape or form is probably as old as the first settled communities, if not older. Neolithic communities probably fermented fruit deliberately and early Egyptian texts refer to its medicinal use. Later, alchemists would distill what we now call ethanol from a fermented mixture to concentrate the "essence" of the mixture. It is no coincidence that, in many parts of the world, we still refer to distilled liquor as "spirits". The alchemists believed that they were separating a metaphysical life-force akin to a spirit from the ferment. Both whiskey and aquavit derive from *Aqua vitae*, meaning water of life.

The philosophy of Vitalism is supposed to have been finally put to bed when Friedrich Wöhler succeeded in synthesizing urea, $(NH_2)_2CO$, from ammonium cyanate, but such paradigm shifts rarely happen overnight. Indeed, the whole homeopathic health industry, still flourishing today, seems to cling to the idea of some mystical insubstantial essence. You'd be surprised to know who still buys into this stuff!

18.5 FROM ORGANIC CHEMISTRY TO BIOCHEMISTRY

Given the infinite number of possible carbon-based compounds, for practical reasons it seems inevitable that life would make a "decision" to use a limited number. There are many variations and modifications that are made, but the universality of a number of

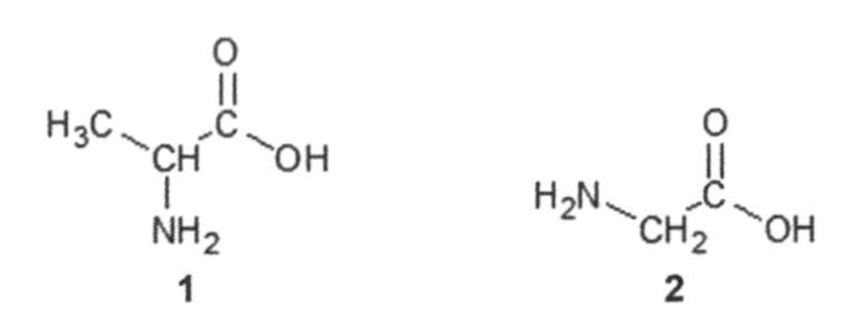

Figure 18.1 Structures of amino acid 1 and 2 (see Q5).

fundamental organic molecules is striking. Five bases are used in RNA and DNA and twenty amino acids are the basis of the vast majority of all proteins. Similar "choices" seem to have been made with monosaccharides and lipids.

The fact that twenty amino acids (albeit with post-translational modifications in many cases) are used to produce the vast array of proteins has been seen to have a striking parallel with the alphabet and language: in English 26 letters give rise to everything from King Lear to Edward Lear! Could a "linguistic" analysis of protein structure prove fruitful in understanding and predicting protein structure? Some have thought so! Predictive engineering of proteins is something of a Holy Grail for molecular biologists, so it's certainly worth a try!

Q5. Two of the twenty amino acids that constitute the basic building blocks of proteins are shown in Figure 18.1.

a) Give the IUPAC name for molecules 1 and 2 in Figure 18.1.
b) Identify the amino acid that can form optical isomers and explain why the other does not.

It didn't take long for scientists to realize that the use of amino acids by living systems was far from random. Only one of the optical isomer variants is used and it has since been suggested that one way to distinguish abiotically synthesized amino acids—such as those in the Miller experiment—from those synthesized by living systems is to look for a significant imbalance between optical isomers. Could such an imbalance be evidence for life?

18.6 *R* OR *S*, + OR

There are two ways to denote optical isomers in the naming system, sadly they can seem confusing. The empirically determined method gives rise to the + and − system. If a solution of one particular optical isomer rotates plane-polarized light in a clockwise direction, it is given the + designation, and − for the anticlockwise direction. The *R* and *S* system is denoted by simple inspection of the molecule and designating on the basis of a priority system, one which I don't propose to go into here. Sadly, there is no simple relationship between *R* and +, and *S* and −, which is why you will see both on reagent bottles.

So, carbon has tremendous versatility, of that there can be no doubt, but the ultimate oxidation product of all carbon-based molecules in a strongly oxidizing environment is just one molecule. Mankind's thirst for chemical energy provides just that environment.In the final section I intend to consider carbon dioxide (CO_2).

18.7 CARBON DIOXIDE AND THE OCEANS

The interaction between carbon dioxide and water is a complicated one, as can be seen from the series of equilibria shown in eqn 18.1–18.4. Carbon dioxide dissolves in water and the position of the first equilibrium is determined by the partial pressure of the carbon dioxide in the air. Once dissolved in water, the diprotic carbonic acid has two pK_a values of 6.38 and 10.32. The position of these last two equilibria is obviously pH dependent and all of them are temperature dependent.

$$1.\ CO_{2(g)} + H_2O_{(l)} \rightleftharpoons H_2O_{(l)} + CO_{2(aq)} \quad (18.1)$$

$$2.\ H_2O_{(l)} + CO_{2(aq)} \rightleftharpoons H_2CO_{3(aq)} \quad (18.2)$$

$$3.\ H_2CO_{3(aq)} \rightleftharpoons H^+{}_{(aq)} + HCO_3{}^-{}_{(aq)} \quad (18.3)$$

$$4.\ H^+{}_{(aq)} + HCO_3{}^-{}_{(aq)} \rightleftharpoons 2H^+{}_{(aq)} + CO_3{}^{2-}{}_{(aq)} \quad (18.4)$$

In addition to these equilibria, any modeling of the interaction between carbon dioxide and the oceans has to take into account precipitation of carbonate ions due to ions in sea water, such as calcium, and the biological impact of photosynthesis.

It's not surprising that modeling something as seemingly simple as the interaction of carbon dioxide with water is complicated, but it is vital! Before we consider the ability of dissolved inorganic carbon (DIC) to help buffer the oceans, let's illustrate how the pH of pure water changes with a single drop of 2 M H_2SO_4.

Q6.

a) What is the pH of 100 cm^3 of pure water at 25 °C?
b) Assuming one drop of 2.00 mol dm^{-3} has a volume of 0.05 cm^3, calculate the number of moles of H^+ ions there are in the drop.

c) After this drop is added to 100 cm^3 of pure water, calculate the pH of the solution.

Hopefully, you will see that the effect on the pH is significant! When you consider that life processes (with some interesting exceptions) need a rather narrow range of pH, the importance of buffering systems becomes apparent. If our own blood pH varies by more than half a pH unit, we are in big trouble!

18.8 BUFFERING THE OCEANS

Q7.

a) Hydrogen carbonate ions play a vital role in buffering the pH of the oceans. Using equations, show how an aqueous solution of hydrogen carbonate ions can buffer the addition of acid and alkali.
b) With reference to eqn 18.1–18.4, describe and explain how an increase in the partial pressure of carbon dioxide in the air will affect the pH of the oceans.

When you hear pronouncements in the media that the pH of the oceans has decreased, on average, by a mere 0.1 of a pH unit you might be forgiven for thinking it's a pretty trivial shift. However, **pH is a logarithmic scale**!

c) Show that a pH shift from 8.2 to 8.1 represents a 26% increase in the H^+ ion concentration.
d) Fig. 18.2 shows the pH range of sea water, which is roughly from 7.9 to 8.5. What impact does a shift towards the acidic end of the range have on the availability of carbonate ions in sea water? How might this impact on organisms that form shells?

18.9 BUFFERING THE BLOOD: CARBONIC ANHYDRASES

If maintaining the pH of the ocean is important then, for animals, maintaining the pH of blood is at least as important. However, with an average human adult capacity of about 5 dm^3 of blood, we have rather less volume capacity to play with than the oceans and, consequently, the system has to respond to changes in pH much more effectively. This is where the enzyme carbonic anhydrase enters the picture or, more correctly, a group of enzymes called carbonic anhydrases. We have already noted that blood pH must remain within a rather narrow range and we have also seen that, without a buffering

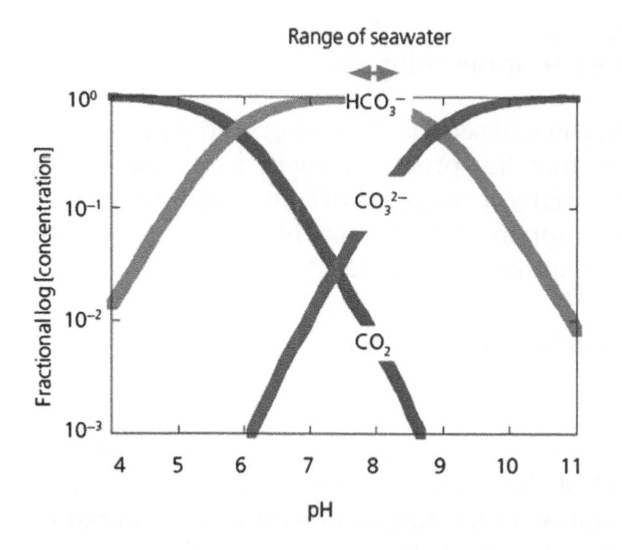

Figure 18.2 Relative proportions of the three inorganic forms of CO_2 dissolved in sea water. Reproduced by kind permission of the Royal Society Report.[1]

system, pH can change significantly with the addition of small quantities of acid or alkali. Carbonic anhydrase is one of the most rapidly functioning enzymes known. Some variants of carbonic anhydrase catalyze the conversion of $CO_{2(aq)} + H_2O_{(l)}$ into $H^+_{(aq)} + HCO_3^-{}_{(aq)}$ at something like 10^6 molecules per second!

Q8.

a) The equilibrium that is catalyzed by carbonic anhydrase is $CO_{2(aq)} + H_2O_{(l)} \rightleftharpoons H^+_{(aq)} + HCO_3^-{}_{(aq)}$. Write an expression for K_a for the forward reaction. (You can assume the $[H_2O]$ to be constant.)

b) If the pK_a for this reaction is 6.38, calculate the equilibrium ratio of $[HCO_3^-]$: $[CO_2]$ under physiological conditions, given that normal blood pH = 7.4.

18.10 CARBON ISOTOPES AND LIFE'S SIGNATURE

Isotopes are atoms of the same chemical element with similar chemical and physical properties. It can no longer be said that they have exactly the same chemical properties and it has been apparent for some time that photosynthesis shows a distinct selective preference for the ^{12}C isotope over the ^{13}C isotope, which has a "natural" abundance of 1.1%. In addition to these stable isotopes of carbon, ^{14}C

is a naturally occurring radioactive isotope with a half life of 5730 years, which decays by a process of β radiation. In the hunt for the first signs of life in some of the Earth's oldest rocks it is this depleted carbon ratio that is used to give evidence for photosynthesis, and thus life. If fossilized structures that resemble cell shapes, *i.e.*, spheres, rods, *etc.* show this depleted ratio then, the reasoning goes, the carbon must have been laid down by a photosynthetic-type process. If, as some studies suggest, life was around 3.4 billion years ago, then it would suggest that life got going pretty quickly after the formation of the Earth, which is thought to be a "mere" 1.2 billion years earlier.

18.11 CONCLUSIONS

In Primo Levi's book *The Periodic Table*[2] the last chapter is titled simply "Carbon". It is a chapter that I read from time to time to remind me that writing about scientific concepts can take your breath away. There are many creation stories—some intimately bound with theologies—that offer grand and impressive accounts about how we came to be here, but I can think of none that are as awe inspiring as that which is produced by the collective works of science. Today, there are scientists who call themselves astrobiologists and astrochemists! It is an acknowledgement that the laws of physics and chemistry are truly universal and that, if they can give rise to life on our planet, there is no reason that they cannot do so elsewhere in the universe. What then are the requirements for this life process to begin? It is difficult to say; some believe that only water is the realistic candidate for life's solvent. Energy is certainly needed, but in what form? There does, however, seem to be a general consensus that only carbon has the versatility to form the basis of life.

REFERENCES

1. Royal Society Report, *Ocean acidification due to increasing atmospheric carbon dioxide*, Policy Document, June 2005, Royal Society, London.
2. P. Levi, *The Periodic Table*, 1986, ABACUS, London.

Answers to questions in this chapter are on pages 224–225.

Answers

CHAPTER 1

A1. Allotropes are different structural forms of the same element. Graphite and diamond are allotropes of carbon, and tin can exist in three allotropic forms: α, β and γ

A2. None of the phosphorus in the body exists as **elemental** phosphorus any more than the sodium content of the body exists as elemental sodium. In the form of chemical compounds, such as calcium phosphate, phosphorus loses its toxic properties.

A3.

a) 14 kg = 14 000 g. 14 000/31 = 451.6 moles of P atoms OR 14 000/124 = 112.9 moles of P_4 molecules

b) $P_4 + 5O_2 \rightarrow P_4O_{10}$.

c) $\Delta H^{\theta}{}_{f}$ for P_4O_{10} is the same as the equation in b). Thus, 112.9 × (−2984) = 336 894 kJ or ≈337 MJ

A4.

a) The structure is analogous to the N≡N molecule:

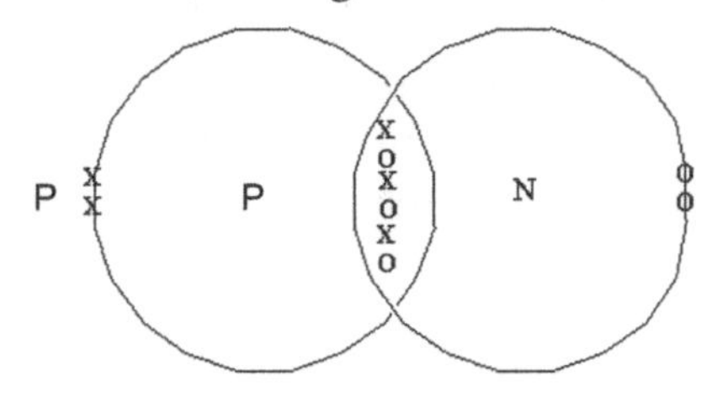

Around the World in 18 Elements
By David A. Scott

Published by the Royal Society of Chemistry, www.rsc.org

b) Remember P_4 forms a tetrahedral, so there are six P–P bonds:

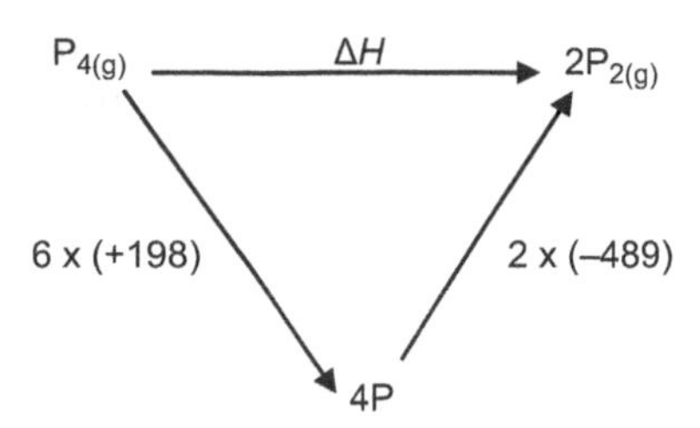

So $\Delta H = 1188 - 978 = +210$ kJ mol^{-1}.

c) Results suggest that P_4 is the more stable allotrope (but see Q11).

A5. P_4S_3 $M_r = 220$ g mol^{-1} and 1 kg = 1000 g, therefore: 1000/220 = 4.545 moles. The ratio is 16 : 3, so (16/3)×4.545 = 24.24 moles of $KClO_3$ needed. M_r $KClO_3 = 122.5$ g mol^{-1}, so 24.24×122.5 = 2969 g or 2.97 kg (to three significant figures).

A6. The correct oxidation numbers are as follows:

a) P_4O_6 $+3$.
b) Ca_3P_2 -3.
c) H_3PO_4 $+5$.
d) Na_2HPO_3 $+3$.

A7. The Al^{3+} ion has a higher charge density than Na^+ due to i) its smaller ionic radius and ii) its higher charge. As a result, it will polarize the electron cloud surrounding the P^{3-} ion.

A8.

a)

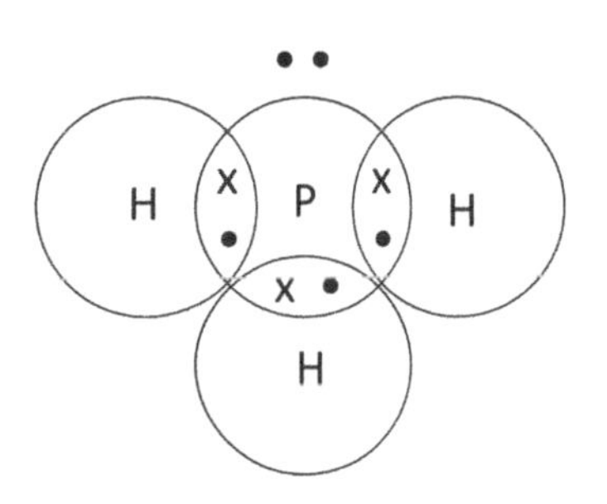

b) Both molecules are pyramidal in shape but, because of the difference in electronegativity between N and H, the NH_3 has a permanent dipole and thus ammonia molecules will form intermolecular hydrogen bonds. These are relatively strong intermolecular forces. There is no permanent dipole in PH_3 because there is no difference in electronegativity between P and H and thus only induced dipole interactions generate relatively weaker forces of attraction between PH_3 molecules.

A9.

a) $1s^2 2s^2 2p^6 3s^2 3p^3$.
b) $PCl_3 + Cl_2 \rightarrow PCl_5$.
c) The $3s^2$ pair of electrons is split and one is promoted into a 3d orbital. There is now one unpaired electron in 3s, three unpaired electrons in each of the 3p x, y and z orbitals and one in a 3d orbital. The five orbitals hybridize to generate the trigonal bipyramidal shape (see part d).
d)

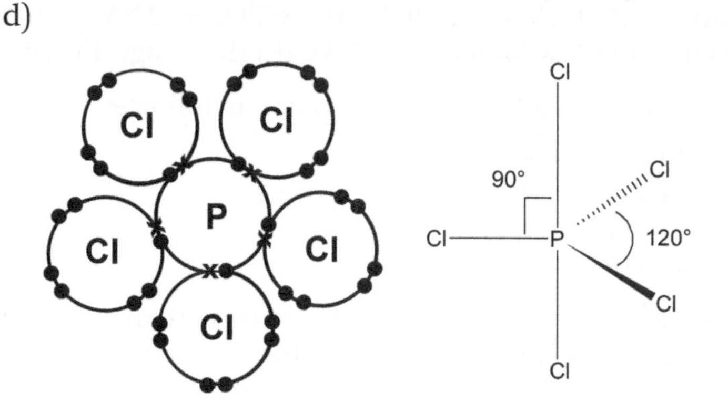

A10.

a)

$$K_a = \frac{[H^+][HPO_4^{2-}]}{[H_2PO_4^-]}.$$

b) pH = 7 is $[H^+] = 1\times10^{-7}$ mol dm^{-3}, so $1\times10^{-7} = K_a \times \frac{[H_2PO_4^-]}{[HPO_4^{2-}]} = \frac{1\times10^{-7}}{6.2\times10^{-8}}$ = ratio of 1.6 : 1 or about 62% $H_2PO_4^-$ to 38% HPO_4^{2-}.

A11. $\Delta S_{system} = \Sigma S_{products} - \Sigma S_{reactants} = (2\times218) - 41 = +395$ J mol^{-1} K^{-1} and $\Delta S_{surroundings} = -\Delta H/T$. Therefore, $\Delta H = \Sigma\Delta H_{products} - \Sigma\Delta H_{reactants} = (2\times144) - 0 = 288$ kJ mol^{-1}.

The reaction becomes feasible when $\Delta S_{system} - \Delta H/T = 0$. Rearranging this equation gives $T = \Delta H/\Delta S_{system}$ and, after converting ΔH into J mol^{-1}, we achieve $T = 288000/395 = 729$ K (456 °C).

A12.

a) Atom economy = (mass of desired product/total mass of products) × 100

$$M_r = \underset{310}{Ca_3(PO_4)_{2(s)}} + \underset{120}{10C_{(s)}} + \underset{360}{6SiO_{2(s)}} \rightarrow 6CaSiO_{3(i)} + 10CO_{(g)} + \underset{124}{P_{4(g)}}$$

Atom economy = (124/790) × 100 = 15.7%.

b) Economic viability could be improved by finding a market for the other products, *e.g.*, $CaSiO_3$ could be used for road surfacing and CO could be used as a fuel (albeit not domestic)!

CHAPTER 2

A1. $(6\times10^{27})\times0.7\times0.55=2.31\times10^{27}$ g.

A2. $Fe=(Ar)\ 3d^6 4s^2$ $Fe^{2+}=(Ar)\ 3d^6$ $Fe^{3+}=(Ar)\ 3d^5$.

A3.

i. For Fe_2O_3: $(2\times56)/(2\times56)+(16\times3)=112/160\times100=70\%$ iron. Oxidation number $=+3$.
ii. For Fe_3O_4: $(3\times56)/(3\times56)+(4\times16)=168/232\times100=72.4\%$ iron. Oxidation number $=+2.6$, although it is actually a mixture of $+3$ and $+2$.
iii. For $FeCO_3$: $56/[(56+12)+(16\times3)]=56/116\times100=48.3\%$. Oxidation number $=+2$.

A4. Numbers: 4, 1 and 1. Disproportionation is the simultaneous oxidation and reduction of the same species. In this case, from $+2$ up to 2.6 and down to 0.

A5.

a) M_r of $(NH_4)_2SO_4\cdot FeSO_4\cdot 6H_2O=(18\times2)+(32+64)+56+32+64+(6\times18)=56/392\times100=14.3\%$.
b) Balanced redox equation: $5Fe^{2+}+MnO_4{}^-+8H^+\rightarrow 5Fe^{3+}+4H_2O+Mn^{2+}$. Reaction ratio is $5Fe^{2+}$: $1MnO_4{}^-$. Moles of Mohr salt used: $23.45/1000\times0.098=2.298\times10^{-3}$ moles of Mohr salt, which contains the same number of Fe^{2+} ions. This will have reduced $2.298\times10^{-3}/5=4.596\times10^{-4}$ moles of $MnO_4{}^-$, which is in 25 cm^3 of solution. Thus, the concentration in the $MnO_4{}^-$ solution $=4.596\times10^{-4}/0.025=0.0184$ M to three significant figures.
c) If a balance has a maximum precision of 0.01 g, then the % error for one measurement for the Mohr's salt is $0.01/392\times100=0.00255\%$ error, but for hydrated iron II sulfate it is $0.01/277.9\times100=0.0036\%$ error.
d) $(NH_4)_2SO_4$ is the salt of a weak base and a strong acid. When this salt dissolves, the $NH_4{}^+$ behaves as a conjugate acid and protonates ^-OH ions.

$$NH_4{}^+ + OH^- \rightleftharpoons NH_3 + H_2O$$

As K_W must remain constant, $[H^+][OH^-]=1\times10^{-14}$, the $[H^+]$ increases. Thus, a solution of Mohr's salt is acidic.

A6. M_r of Fe^{2+} in the porphyrin ring = 552 g mol^{-1}.

A7. 1650 atoms of Fe.

A8. $\frac{1}{2}O_2 + 2H^+ + 2e^- \rightleftharpoons H_2O \quad E^\theta = +1.23$
and
$2Fe^{2+} \rightleftharpoons Fe^{2+} + 2e \quad E^\theta = -0.77$.
Therefore, E^θ cell = 1.23 – 0.77 = +0.46 e/V, so the reaction is feasible.

A9.

a) Moles of $KMnO_4$ used = 29.75/1000×0.02 = 5.95×10^{-4}. Thus, 5×(5.95×10^{-4}) of Fe^{2+} oxidized = 2.975×10^{-3}. This is in 25 cm^3, so (2.975×10^{-3})×20 = 0.0595 in the original 500 cm^3 from which came the 3.50 g of steel.
0.0595 moles of Fe has a mass of 0.0595×55.8 = 3.32 g. Thus, %iron = (3.32/3.50)×100 = 94.9% iron.

b) Assumptions include:
- Iron is the only species in the alloy that can be oxidized.
- Any Mn^{2+} ions could be oxidized and give an overestimation of iron.
- None of the iron has been oxidized by the acid beyond Fe^{2+}, which would give an underestimation of the iron content.

A10.

a) A solution of Fe^{3+} ions is sufficiently acidic to be appreciably dissociated. pK_a = 2.2; thus, K_a = 6.0×10^{-3}. 6.0×10^{-3} = $\alpha^2/(0.05 - \alpha)$. This requires solving a quadratic equation, giving a value for α = 0.0146 equivalent to a pH of 1.84 to two decimal places.
Note that, in this case, assuming $(0.05 - \alpha) \approx 0.05$ would give a pH value of 1.76, which is appreciably different to the value calculated. However, a second pK_a to consider is $[Fe(H_2O)_5OH]^{2+} + H_2O \rightleftharpoons [Fe(H_2O)_4(OH)_2]^+ + H_3O^+$, and thus the pH may be lower.

b) Iron III carbonate does not exists because iron III salts (as we have seen) are acidic in solution and thus carbonate would be forced from solution:

$$2H^+ + CO_3{}^{2-} \rightarrow H_2O + CO_2$$

CHAPTER 3

A1.

a) Bonds broken $945 + 496 = 1441$. Bonds formed $2 \times (-631) = -1262$. Net enthalpy change $= +179$. So, $\Delta H_R = +179$ kJ.
b) Entropy change $= (2 \times 211) - (103 + 96) = 222$ J K^{-1}.
c) When $\Delta H - T\Delta S = 0$, then the reaction becomes feasible. So using a) and b) $179 - (T \times 222/1000) = 0$.
d) Solving for $T = 806.3$ K or 533.3 °C.
e) The fact that the reaction becomes thermodynamically feasible at this temperature tells us nothing about reaction rate.

A2.

a) If it was a one-step reaction, there would need to be a simultaneous collision of two NO molecules and one O_2 molecule with the minimum activation energy and possibly the correct orientation of all three molecules. This is, statistically, extremely unlikely!
b)
  (i) $x = 2$; $y = 1$.
  (ii) $k = \text{Rate}/[NO]^2[O_2]$. Inputting values from experiment one, $k = 0.0222/(0.0222)^2(0.00866) = 5201.5$ dm^6 mol^{-2} s^{-1}.
  (iii) A suggested mechanism for the reaction is as follows: step 1 is fast and involves both NO molecules forming an intermediate dimer. Step 2 involves the reaction between this dimer and O_2 to form products. If this second step is the rate determining step, it explains why both the rate equation contains the $[NO]^2$ factor (which determines the formation of N_2O_2 and the $[O_2]$, which then reacts to form products):

$$\text{Step 1: } 2NO \rightarrow N_2O_2 \text{ (fast)}$$

$$\text{Step 2: } N_2O_2 + O_2 \rightarrow 2NO_2 \text{ (rate determining)}$$

Why does an increase in temperature **decrease** reaction rate? It is thought that an increase in temperature makes the N_2O_2 less stable (the increasing kinetic energy with temperature causes the dimer to revert to two NO molecules). This is consistent with a slowing in reaction rate as T increases.

A3. Nitrogen is assigned an oxidation number of $+4$ in both NO_2 and N_2O_4.

A4. This is a disproportionation reaction in which nitrogen in oxidation state $+2$ (NO) is being both oxidized ($+4$ in NO_2) and reduced ($+1$ in N_2O). As the reaction proceeds from left to right the gaseous mixture will turn brown, as NO_2 is brown, but the other two oxides are colourless.

A5.

a) $Cu + 4H^+ + 2NO_3^- \rightarrow Cu^{2+} + 2NO_{2(g)} + 2H_2O$.
b) $3Cu + 8H^+ + 2NO_3^- \rightarrow 3Cu^{2+} + 2NO_{(g)} + 4H_2O$.
c) Given that both gases are produced when copper reacts with concentrated nitric acid, collecting the gases over water would remove most of the soluble NO_2 allowing the NO to be collected.

A6. Comproportionation is essentially the reverse of disproportionation. When the same element in two different oxidation states are simultaneously reduced/oxidized. In this case the nitrogen in NH_4^+ is oxidized from -3 to $+1$, whilst the nitrogen in NO_3^- is reduced from $+4$ to $+1$.

A7. Aerobic bacteria require oxygen. *Nitrosospira* obtain their oxygen from nitrate and thus, if the ^{15}N isotope is labeled on the nitrate group, one might expect to find the isotope present in the $^{15}N_2O$ produced by the bacteria. Anaerobic mechanisms would utilize the nitrogen from the ammonium species. Thus, providing a nutrient with the ^{15}N isotope present in either nitrate or ammonium would help determine whether the isotope came from either the NO_3^- or the NH_4^+. Although this is almost certainly a simplification it might give a preliminary indication of what is taking place.

A8. Assuming room temperature/pressure (1 mole of gas occupies 24 dm^3), then 15 $dm^3 = 15/24 = 0.625$ moles. Nitride to nitrogen ratio$= 2:3$, so $(2/3 \times 0.625) = 0.416$ moles of NaN_3 needed. The mass is (0.416 x $(23 + (14 \times 3) = 27.04$ g.

In reality a pressure greater than atmospheric pressure inside the bag would be needed if the air bag is to be effective.

A9. Empirical: $C_3H_7N_2O$; Molecular $= C_6H_{14}N_4O_2$.

A10. Percentage nitrogen: $(14 + 14)/((12 \times 3) + (7) + 28 + 16) \times 100 =$ 32.2% to 1 d.p.

A11. pK_a for $R\text{-}NH_3^+$ means that the species is very weakly dissociated to $R\text{-}NH_2 + H^+$. In a pH of 7.4 the $[H^+]$ ion concentration will shift the equilibrium even further back to the protonated form, making it effectively 100% protonated.

A12. 6 moles of N≡N bonds $= 6 \times 945 = 5670$ kJ. This would be released as heat energy!

A13. 2,4,6-Trinitromethyl benzene.

A14. $K_p = (NH_3)^2/(N_2) \times (H_2)^3$. So, $1.45 \times 10^{-3} = p(NH_3)^2/(0.432) \cdot (0.928)^3$; therefore, $p(NH_3)^2 = 1.45 \times 10^{-3}/0.345$, giving $p(NH_3) = 0.0648$ atm.
Partial pressure $= 0.0648$.
Mole fraction = partial pressure/total pressure $= 6.48 \times 10^{-2}/(0.432 + 0.928 + 6.48 \times 10^{-2}) = 0.0456$.
Percentage = mole fraction $\times 100 = 4.56\%$.

A15. Oxidation states for each species are:

$$\underset{-3}{4NH_3} + 5O_2 \rightarrow \underset{+2}{4NO} + 6H_2O$$

$$\underset{+2}{2NO} + O_2 \rightarrow \underset{+4}{2NO_2}$$

$$\underset{+4}{4NO_2} + O_2 + 2H_2O \rightarrow \underset{+5}{4HNO_3}.$$

CHAPTER 4

A1. $SO_2 = +4$; $SO_3 = +6$; $H_2S = -2$; $(CH_3)_2S = -2$; $(CH_3)_2SO = 0$; $FeS = -2$; $FeS_2 = -1$; $CaSO_4 \cdot 2H_2O = +6$.

A2. 1.06×10^8 tonnes $= 1.06\times10^{14}$ g. Moles of $SO_2 = (1.06\times10^{14})/64 = 1$ moles. One mole at room temperature/pressure occupies $24\,dm^3 = 2.4\times10^{-2}\,m^3 = 2.4\times10^{-8}\,km^3$. Thus, 1.453×10^{12} moles occupies $(1.656\times10^{12})\times(2.4\times10^{-8}) =$ 39 744 km^3 sufficient to cover Wales 1.91 times to a depth of 1 km.

A3.

a)

Ore	*Formula*	*% Sulfur*
Molybdenite	MoS_2	40%
Pyrite	FeS_2	53%
Sphalerite	ZnS	67%
Galena	PbS	13%
Chalcopyrite	$CuFeS_2$	35%

b) You would need to consider i) grade of ore, ii) accessibility of ore and iii) potential market for other products from extraction. None of these ores would, in fact, be mined for the sulfur. In each case the metals are more valuable. Most sulfur is obtained from petrochemical desulfurization processes.

A4.

a)

$$\underset{-2}{2H_2S} + 3O_2 \rightarrow \underset{+4}{2SO_2} + 2H_2O$$

$$\text{then } \underset{-2}{2H_2S} + \underset{+4}{SO_2} \rightarrow \underset{0}{3S} + 2H_2O.$$

b) Combining the equations gives a $4H_2S : 3S$ ratio. Thus, $500\,dm^3$ H_2S is $500/24 = 20.83$ moles, so $\frac{3}{4}\times20.83 = 15.625\times32 = 500$ g.

A5.

a) Standard entropy change $((256.8\times2) - (248.1\times2) - 102.5) = 513.6 - 598.7 = -85.1\,J\,K^{-1}$.

b) Total entropy change $= \Delta S_{\text{system}} - \Delta H/T$.
So, the total entropy change at 25 °C $= -85.1 - (-197\,000)/298 = +576$ J K^{-1}.

c) At 400 °C (673 K), the total entropy change is now $-85.1 - (-197\,000)/673 = +207.6$ J K^{-1}. At this temperature the rate of reaction means that the position of equilibrium is reached quickly and the total increase in entropy is still sufficiently large to favour the forward reaction.

d) $\Delta S_{\text{Total}} = R\ln K$. So, $\ln K = +207.6/8.314 = 24.97$; so, $K = e^{24.97} = 6.99 \times 10^{10}$ atm^{-1}.

e) The overall equation is

$$2SO_2 + O_2 \rightleftharpoons 2SO_3$$

f) Vanadium is reduced from +5 to +4 in the first step and the oxidised from +4 back to +5 in the second step thus regenerating the catalyst.

A6.

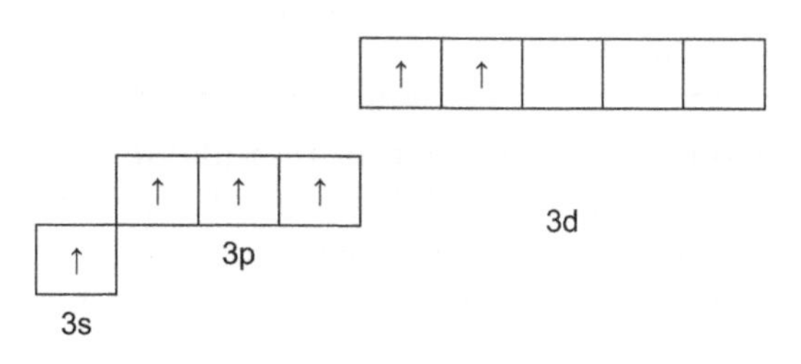

The 3s and 3p paired electrons are promoted to the 3d orbitals. Thus, sp^3d^2 hybridization allows the formation of six covalent bonds.

A7.

a) A conjugate base is the H^+ ion acceptor for the reverse right to left reaction as the equilibrium is written.

b)

$$H_2O_{(l)} + SO_4{}^{2-}{}_{(aq)} \rightleftharpoons HSO_4{}^{-}{}_{(aq)} + OH^{-}{}_{(aq)}$$

For aqueous solutions, $pK_a + pK_b = 14$, thus $pK_b = 12$; so, $K_b = 1 \times 10^{-12}$ mol dm^{-3}.

A8.

a) A mass of 1.21 g of $BaSO_4$ precipitated. Thus, 1.21/233 $= 5.19 \times 10^{-3}$ moles of $SO_4{}^{2-}$ ions in 25 cm^3. Therefore, there are 5.19×10^{-2} moles in 250 cm^3, which has a mass of $5.19\times10^{-2}\times(M_R$ of $CaSO_4\cdot2H_2O)=8.927$ g; so, percentage $=8.927/10.30\times100=86.7\%$ purity.

Storage conditions are important as the sample may either absorb or lose water from the air depending on ambient humidity.

b) K_{sp} for barium sulfate $=[Ba^{2+}][SO_4^{2-}]=1\times10^{-10}$ mol^2 dm^{-6}, which means that $[Ba^{2+}]$ in solution $=1\times10^{-5}$ M. One mole of $BaSO_4$ dissolves for every mole of Ba^{2+}, thus $1\times10^{-5}\times(233)=2.33\times10^{-3}$ g per dm^3 or 0.1165 g in 50 dm^3.

Thus, Ba^{2+} ions may be toxic but $BaSO_4$ is so insoluble that hardly any will be absorbed from the intestine.

A9.

a)

$$Ca(HSO_3)_2 \rightleftharpoons Ca^{2+}{}_{(aq)} + 2HSO_3^-{}_{(aq)}$$
$$+4$$
$$HSO_3^-{}_{(aq)} \rightleftharpoons H^+{}_{(aq)} + SO_4^{2-}{}_{(aq)}.$$

b) $pK_a=7.2$, $K_a=6.31\times10^{-8}$. 0.01 M solution of $Ca(HSO_3)_2=0.02$ M with regards to HSO_4^-. Thus, $[H^+]=\sqrt{6.31\times10^{-8}/0.02}=1.776\times10^{-4}$. So, pH $=3.75$.

A10.

a) M_R of $Na_2S_2O_3=158$. If the weighing balance has an accuracy of ±0.01, then the error for sodium thiosulfate is $0.01/158=\pm6.3\times10^{-5}$ mole. If M_R is smaller, *e.g.*, 40 g mol^{-1}, then the error is $\pm0.01/40=2.5\times10^{-4}$ moles, which is a greater error.

b)

$$2S_2O_3^{2-} \rightarrow S_4O_6^{2-} + 2e^- \quad -0.09$$

$$I_2 + 2e \rightarrow 2I^- \quad +0.54$$

$$2S_2O_3^{2-} + I_2 \rightarrow S_4O_6^{2-} + 2I^- \quad +0.45$$

A positive $E^{\theta}{}_{CELL}$ value means the reaction is feasible.

c) 50 cm^3 of a 0.003 M I_2 solution is $50/1000\times0.003=1.5\times10^{-4}$ moles, which is reduced by $2\times(1.5\times10^{-4})$ moles of thiosulfate $=3.0\times10^{-4}$ moles in $2\times10^{-3}=3.0\times10^{-4}$ V^{-1}; so, V $=0.15$ dm^3

A starch indicator is needed to determine the end point.

A11. Acidic conditions are likely to protonate the primary amine group, thus making it more soluble in water due to favourable electrostatic interactions with water molecules.

A12. EI mass spectrometry ionizes the molecules by knocking electrons off the molecule. Larger molecules are likely to be fragmented by this process and so molecular ion peaks will be very rare. However, an m/z peak of 92 is likely to represent a more stable ion fragment, in this case the $[C_6H_4NH_2]^+$ ion.

A13.

a) Removal of electrons is oxidation. The bond angle is based on a distorted tetrahedral, similar to a water molecule; thus, it is approximately 104–105°.
b) The lone pairs on sulfur atoms can act as electron pair donors.

A14. $H_2S + 4H_2O \rightarrow SO_4^{2-} + 8e^- + 10H^+$.

A15.

a) ${}^{35}_{16}S \rightarrow {}^{35}_{17}CI + \beta$. Chlorine is produced.
b) $kt_{\frac{1}{2}} = 0.693$; so, $k = 0.683/44.3 = 0.0156$ days^{-1}. So, $\ln c - \ln c_0 = -kt$, ln(1) – In(10) = –0.0156 × t, $t = 147.6$ days.

CHAPTER 5

A1.

a) It is the energy released when one mole of substance is formed from its constituent gaseous ions. The products must be in their standard states under standard conditions.

b)

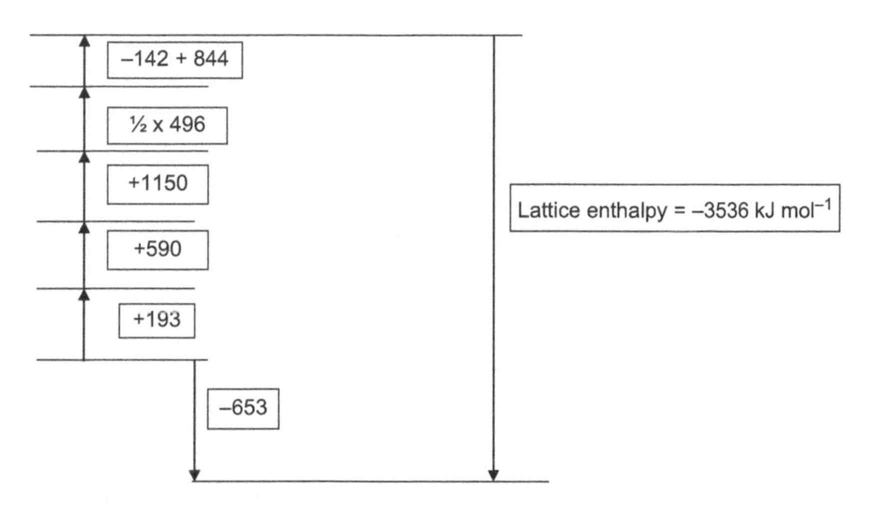

c) After adding one electron, the oxygen has a single negative charge. In order to complete the outer shell a second electron must be added, which involves overcoming electrostatic forces of repulsion between the electron and the O^- ion. This energy is more than compensated for by the extremely exothermic lattice enthalpy of CaO.

d) The theoretical value is based on 100% covalent character. The discrepancy would suggest a small degree of covalent character in the CaO structure.

A2.

a) The bonding in $CaCl_2$ is an ionic lattice.

b) Cathode: $Ca^{2+} + 2e^- \rightarrow Ca$.
Anode: $O^{2-} \rightarrow \frac{1}{2}O_2 + 2e^-$.

c) $Q = I \times t$ (t = time in s). Total charge, $C = 15\,000 \times 8 \times 60 \times 60 = 4.32 \times 10^8$ C. This is $4.32 \times 10^8 / 96480 = 4477.6$ moles of electrons, which will deposit $\frac{1}{2} \times 4477.6 = 2238.8$ moles of Ca, which has a mass of 89 552.24 g, or 90 kg of Ca to the nearest kg.

d) Hazards are inherent properties of chemical substances that we cannot alter; *e.g.*, the reactive nature of calcium with air, water *etc.* Risk is assessed in terms of the likelihood of an accident once precautionary measures have been taken. Risk can be minimised by making adjustments to the manufacturing procedures.

A3.

a) Wavelength $= 616$ nm or 6.16×10^{-7} m. Frequency $= c/\lambda = 3.0 \times 10^{8}/6.16 \times 10^{-7} = 4.87 \times 10^{14}$ Hz (or s^{-1}).
b) Energy $= 6.63 \times 10^{34} \times 4.87 \times 10^{14} = 3.23 \times 10^{-19}$ J.

A4. 70% CaO means 700 kg in 1 tonne and 2% $CaSO_4 \cdot 2H_2O = 20$ kg. Therefore, $(40/56 \times 700) + (40/172 \times 20) = 500 + 4.65 = 504.65$ kg of Ca.

A5.

a) Heat energy is transferred to both the calorimeter and water. The two values must be calculated separately.
$\Delta T = (22.5 - 16.5) = 6$.
Energy transferred to copper calorimeter $= 50 \times 0.4 \times 6 = 120$ J.
Energy transferred to water $= 100 \times 4.2 \times 6 = 2520$ J.
Total energy transferred to calorimeter and water $= (2520 + 120) = 2640$ J for 14.8 g of Ca_3SiO_5.
M_R $Ca_3SiO_5 = 228$ g mol^{-1}, so $14.8/228 = 0.065$ moles.
Thus, $2640/0.065 = 40615.4$ J, so $\Delta H_R = -40.6$ kJ mol^{-1} to 3 significant figures.
b) The discrepancy between -87 and -40.6 kJ mol^{-1} will be largely due to the cement itself, which retains a considerable proportion of the heat energy.

A6. $Ca(HCO_3)_{2(aq)} \rightarrow CaCO_{3(s)} + H_2O_{(l)} + CO_{2(g)}$
160 g $dm^{-3} = 160/102 = 1.57$ mo ldm^{-3}. Thus, 5 dm^3 will contain 7.84 moles. The ratio is 1 : 1, so if all calcium hydrogen carbonate decomposes then 7.84 moles of $CaCO_3$ will be deposited, which is $(7.84 \times 100) = 784$ g.

A7.

a) The hydrocarbon chain is hydrophobic, which means water repelling. The carboxylate group will form favorrable electrostatic interactions with water and this is hydrophilic.
b) $C_{17}H_{35}O^{2-}Na^{+}$ $M_R = 294$ g mol^{-1}.

A8.

a) 294 mg dm^{-3}.
b) Hexadentate meaning, literally, "six teeth". The EDTA forms six covalent bonds with Ca^{2+}.
c) They are co-ordinate or dative bonds in which the EDTA is the electron pair donor.

A9. Approximately 60 inside and 600 000 outside (for an equal volume of solution)!

A10.

a) $K_{sp} = Ca_3(PO_4)_2$ at 25 °C $= 2.07 \times 10^{-33}$ mol^5 dm^{-15}
$[Ca^{2+}]^3[PO_4{}^{3-}] = 2.07 \times 10^{-33}$, $\sqrt[5]{2.07} \times 10^{-33} = 2.91 \times 10^{-7} \times 310 = 9.021 \times 10^{-5}$ g dm^3 or 90.21 μg dm^{-3}.
b) Calcium phosphate has such a low solubility that any free calcium ions would precipitate out as phosphate, making it unavailable for reactions in aqueous solution.

A11.

a) 2-Amino butanedioic acid.
b) One of the lone pairs of the OH group can act as an electron pair donor.
c) At higher pH the carboxylic acid group is likely to be deprotonated, giving the carboxylate groups a net negative charge. Thus, there will also be an electrostatic component to the attractive force between the Ca^{2+} ion and the amino acid.

CHAPTER 6

A1.

a) $(1.4/100)\times 204+(24.1/100\times 206)+(22.2/100\times 207)+(52.3/100\times 208)=207.24$.
b) 4.14×10^{16} atoms.
c) 14.25 μg.

A2.

a)

$$\underset{+2\,-2}{2PbS} + 3O_2 \rightarrow \underset{+2}{2PbO} + \underset{+4}{SO_2}$$

$$\text{then}\quad \underset{+2}{2PbO} + \underset{+2\,-2}{PbS} \rightarrow \underset{0}{3Pb} + \underset{+4}{SO_2}.$$

The second step is a redox reaction.
b) The overall reaction is $3PbS+3O_2\rightarrow 3Pb+3SO_2$. Therefore, the atom economy = (mass of Pb/Total mass of reactants)×100 = (621/813)×100 = 76.4%.
c) 500 kg of Pb is $5\times 10^5/207=2415.5$ moles of Pb. The would come from and equal % of PbS which has a mass of $2415.5\times 239=5.773\times 10^5$ g of PbS (577.3 kg) thus % galena in ore is (577.3×100)/850 = 67.9%.

A3.

a) $Pb(s)+\frac{1}{2}O_2\rightarrow PbO(s)$. The standard enthalpy of formation is the enthalpy change when one mole of products is formed from its constituent atoms. All reactants and products must be in their standard states.
b)

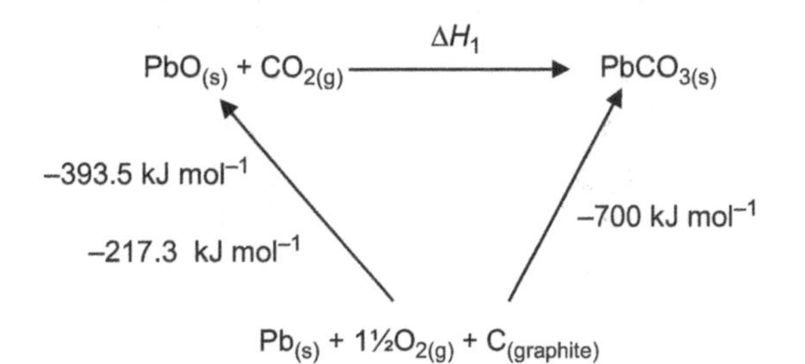

By Hess's law, $\Delta H_1=(-393.5)+(-217.3)+(+700)=+89.2$ kJ mol^{-1}.

A4.

a) 0.3 ppm equal to 0.3 mg dm^{-3}. So, $0.3/207 = 1.45\times10^{-3}$ M.
b) $CaCO_3$ $K_{sp} = 6.0\times10^{-9}$ mol^2 dm^{-6}, so $[CO_3{}^{2-}] = \sqrt{6.0\times10^9} = 7.75\times10^{-5}$ M.
c) K_{sp} for $PbCO_3 = 7.4\times10^{-14}$ mol^2 dm^{-6}, so in a solution where $[CO_3{}^{2-}] = 7.75\times10^{-5}$, the max $[Pb^{2+}] = 7.4\times10^{-14}/7.75\times10^{-5} = 9.55\times10^{-10}$ M. This is $9.55\times10^{-10}\times207 = 1.98\times10^{-7}$ ppm.
d) The presence of carbonate in water reduces the concentration of lead ions by a factor of about 1.5 million! Lead pipes in hard water areas are safe.

A5.

a) 17.13 g of $Pb_3O_4 = 17.13/685 = 2.5\times10^{-2}$ moles. Eqn 6.4 shows a 1 : 1 ratio, so the same number of moles of PbO_2 could theoretically be collected. This is $2.5\times10^{-2}\times(239) = 5.975$ g. If 88% is recovered, then $(88/100)\times5.975 = 5.26$ g.
b) Pb_3O_4 is a mixed oxide. Thus, the reaction is not a redox reaction as both +2 and +4 oxidation states of Pb are present in Pb_3O_4.

A6.

a) A secondary cell is one that can be recharged by reversing the current, thus converting electrical energy back to chemical energy.
b)

$$Pb_{(s)} \rightarrow Pb^{2+}{}_{(aq)} + 2e^- \quad E^\theta = +0.13 \text{ eV}$$

$$PbO_{2(s)} + 4H^+_{(aq)} + 2e^- \rightarrow Pb^{2+}_{(aq)} + 4H_2O_{(l)} \quad E^\theta = +1.47 \text{ eV}$$

$$PbO_{2(s)} + 4H^+_{(aq)} + Pb_{(s)} \rightarrow 2Pb^{2+}_{(aq)} + 4H_2O_{(l)} \quad E^\theta_{CELL} = +1.60 \text{ eV}$$

A7. ${}^{210}_{82}Pb \rightarrow {}^{210}_{83}Bi + \beta$. Bismuth is the product.

A8.

a) $C_8H_{20}Pb$.
b) Tetrahedral (109.5°) bond angle. Pb in same group as carbon and it is probable that it has sp^3 hybridization.

A9.

a) Approximately 0.8 μmol dm^{-3}.
b) 0.166 ppm (*i.e.*, mg dm^{-3} = ppm for solutions).

A10.

a)

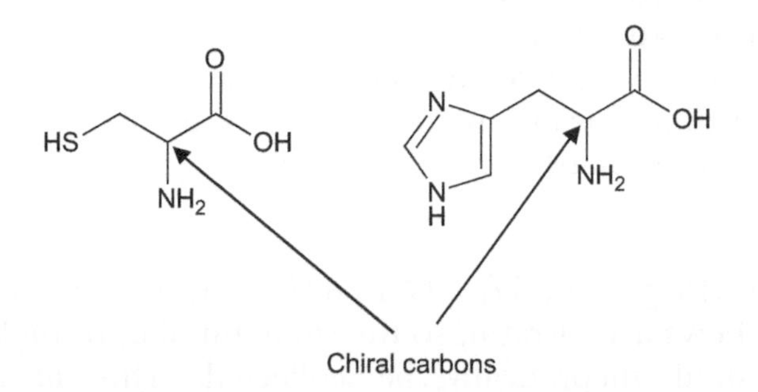

b) Histidine has π electrons and $\pi \rightarrow \pi^*$ electron transitions give rise to absorbances in the UV spectrum.
c) Any lone pairs of electrons are capable of forming dative covalent bonds with Pb^{2+} ions, *i.e.*, the sulphur in cysteine and nitrogen in histidine.

CHAPTER 7

A1.

a) $Li = 1s^2 2s^1$; $Mg = 1s^2 2s^2 2p^6 3s^2$.

b)

(i) $Li_{(g)} \rightarrow Li^{+}{}_{(g)} + e^{-}$.

(ii) $Mg^{+}{}_{(g)} \rightarrow Mg^{2+}{}_{(g)} + e^{-}$.

c) Although Mg has an extra shell, it loses both 3s electrons to form Mg^{2+} ions in the metallic crystal. The nuclear charge of +12 draws the remaining ten electrons closer to the nucleus. Lithium loses its 2s electron but, with a nuclear charge of +3, will "contract" the electronic radius to a lesser degree.

d) Other diagonal relationships include Be and Al, and B and Si.

A2. $E = hf$ and $c = f\lambda$.

$\lambda = 670.8$ nm $= 6.708 \times 10^{-7}$ m. This gives: $f = 3.0 \times 10^8 / 6.708 \times 10^{-7} = 4.47 \times 10^{14}$ Hz. Thus, one photon energy is $(6.63 \times 10^{-34}) \times (4.47 \times 10^{14}) = 2.965 \times 10^{-19}$ J.

A3.

a) Let x = isotope ^6Li fraction, which means $(100 - x)$ = isotope ^7Li fraction.

So $((x/100) \times 6) + ((100 - x)/100 \times 7) = 6.94$, solving for $x = 6\%$ isotope ^6Li, which leaves 94% isotope ^7Li.

b) No, they have the same line emission spectra as line emissions are due to electronic transitions.

A4. $-1 \times 96\,500 \times 2.71 = -240\,000 - (298 \times \Delta S)$. $\Delta S = +72.2$ J K^{-1} mol^{-1}.

A5.

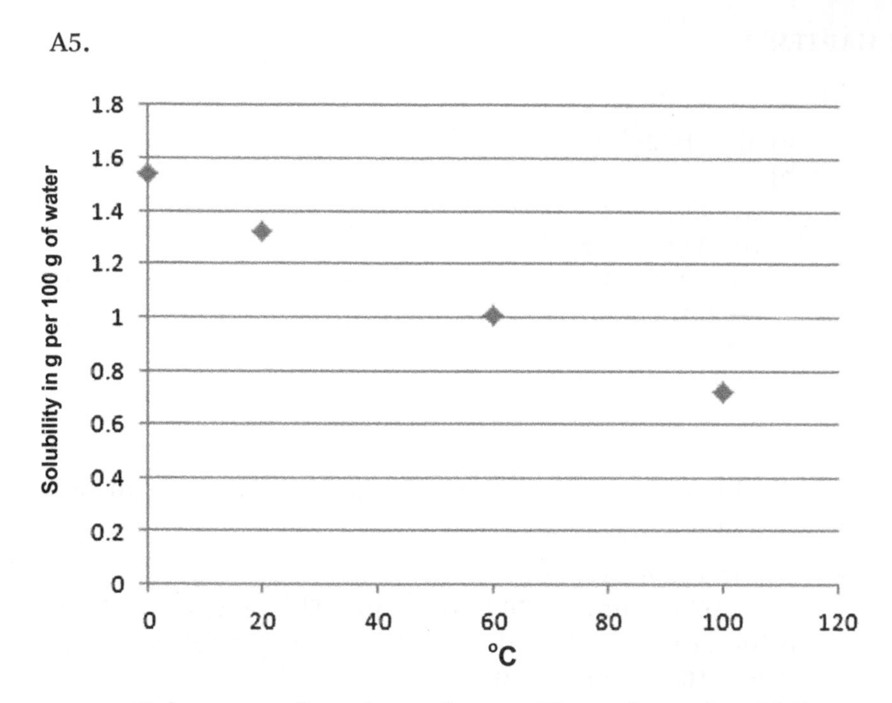

Points are plotted as shown. The points should be connected by a line of best fit and a line extended from the *x* axis at 40 °C to the line and drawn across to the *y* axis gives the solubility value. The value should be in the region of 1.15 g.

A6.

a) $2LiNO_3 \rightarrow Li_2O + 2NO_2 + O_2$.
   $2LiCO_3 \rightarrow Li_2O + 2CO_2$.
b) The higher charge density of the Li^+ ion means that it has a greater polarizing power than the other group I ions and is similar to the charge density of group II +2 ions.

A7.

a) $4LiH + AlCl_3 \rightarrow LiAlH_4 + 3LiCl$.
b) Atom economy = $[(7+27+4)/(7+27+4)+(21+106.5)] \times 100 = 23\%$ atom economy to 1 significant figure.
c) 10 kg of LiH = 10 000/8 = 1250 moles of LiH. Thus, 1250/4 = 312.5 moles of $AlCl_3$ needed. Mass = $(312.5 \times 133.5) = 41719$ g or 42 kg.

d)

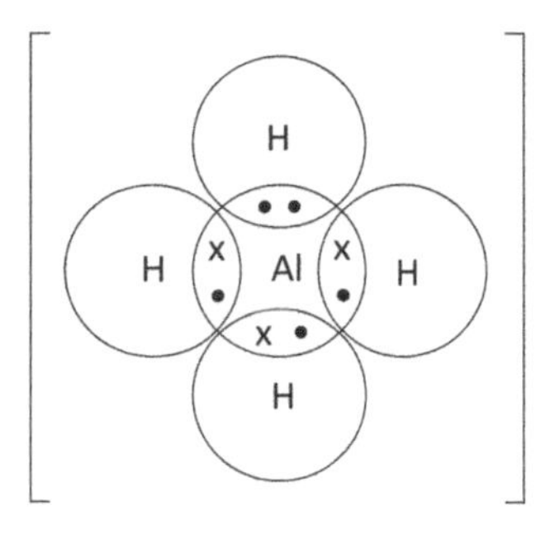

A8.

a) $LiAlH_4 + 4H_2O \rightarrow Al(OH)_3 + LiOH + 4H_2$.
b) 1 $m^3 = 1000$ $dm^3 = 1000/24 = 41.6$ moles of hydrogen at room temperature/pressure. A ratio of 1:4 means $41.6/4 = 10.416$ moles of $LiAlH_4$ needed, which has a mass of $10.416 \times 38 = 395.81$ g to 2 d.p.
c) Any exposure to moist air would result in the gradual decomposition of $LiAlH_4$.

A9.

a)

Organic reactant formula	*Name of reactant*	*Product formula after reacting with $LiAlH_4$ in ethoxyethane*	*Name of product*
$CH_3CH_2CH_2CH_2CH_2Cl$	1-Chloropentane	$CH_3CH_2CH_2CH_2CH_3$	Pentane
CH_3CH_2CHO	Propanal	$CH_3CH_2CH_2OH$	Propan-1-ol
$(C_2H_5)_2CO$	Pentan-3-one	$CH_3CH_2CH(OH)CH_2CH_3$	Pentan-3-ol
$CH_3CH_2CONH_2$	Propanamide	$CH_3CH_2CH_2NH_2$	Propylamine
CH_3CH_2COOH	Propanoic acid	$CH_3CH_2CH_2OH$	Propane-1-ol
CH_3COCH_2COOH	2-Oxopropanoic acid	$CH_3CH(OH)CH_2CH_2OH$	Butane-1,3-,diol

b) The chiral carbon is marked with an asterisk (*):

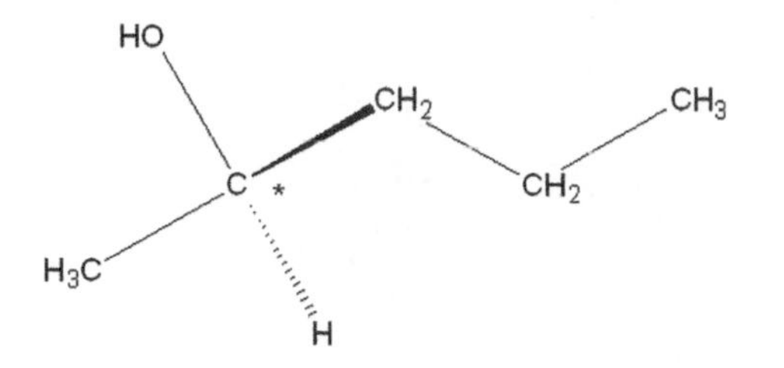

c) As the carbonyl C=O bond is reduced, the specific absorbance of this bond (1650–1750 cm^{-1}) would gradually disappear and the absorbance of the OH bond at a higher wavenumber (3000 cm^{-1}) and above would increase.

A10.

a) i) $2NaOH + CO_2 \rightarrow Na_2CO_3 + H_2O$.
   ii) $2LiOH + CO_2 \rightarrow Li_2CO_3 + H_2O$.
b) $50/24 = 2.083 \times 2 = 4.16$ moles of each hydroxide needed.
c) The molar mass ratio = 40 : 24 = 1.6 : 1, so we need 1.6 kg of NaOH for every 1 kg of LiOH to absorb the same volume of CO_2.

A11.

a) $Li_2CO_{3(s)} + 2HCl_{(aq)} \rightarrow 2LiCl_{(aq)} + CO_{2(g)} + H_2O_{(l)}$.
b) 2.5 mmol dm^{-3} is $2.5/1000 \times 7 = 0.0175$ g in 1 dm^3 or 0.00175 g per 100 cm^3 = 1.75 mg per 100 cm^3.

CHAPTER 8

A1.
$$\underset{-1}{8I^-} + 10H^+ + \underset{+6}{SO_4^{2-}} \rightarrow \underset{-2}{H_2S} + \underset{0}{4I_2} + 4H_2O.$$

Sulfate VI ions are the oxidizing agents and iodide ions are the reducing agents.

A2. A Lewis base is an electron pair donor. The I_3^- ion is linear. The I_2 molecule has 5d orbitals, which can accept an electron pair.

A3.
$$2H^+ + H_2O_2 + \cancel{2e^-} \rightarrow 2H_2O$$
$$2I^- \rightarrow I_2 + \cancel{2e^-}$$
$$\overline{2H^+ + H_2O_2 + 2I^- \rightarrow I_2 + 2H_2O}$$

A4. 10.05 cm^3 of 0.001 M $Na_2S_2O_{3(aq)}$ contains 1.005×10^{-5} moles of thiosulfate, which has reduced $\frac{1}{2}\times1.005\times10^{-5}=5.025\times10^{-6}$ moles of I_2. Scaling up $\times20$ for the total I_2 in 500 $cm^3=1.005\times10^{-4}$ moles of I_2.

This has a mass of $1.005\times10^{-4}\times254=0.0255$ g of I_2 in 10 g of sample$=0.255\%$ by mass.

A5.

a) Molecular formula $C_{15}H_{11}NO_4I_4$ $M_R=777$ g mol^{-1}. Percentage iodine$=508/777\times100=65.4\%$ by mass.
b) The dashed line represents hydrogen bonded to a chiral carbon. This tells us that there are two optical isomers about this carbon, only one of which may be active.
c) At pH$=3$, you might expect the NH_2 group to be protonated, giving the NH_3^+ group. At pH$=11$, the carboxylic acid group would be deprotonated, giving the COO^- carboxylate group.

A6. 4.85 cm^3 of 0.002 M $Na_2S_2O_3=9.7\times10^{-6}$ moles of thiosulfate equivalent to 4.85×10^{-6} moles of iodine. Using eqn 8.6, this converts to $4.85\times10^{-6}/3=1.617\times10^{-6}$ moles of iodate V in a 50 cm^3 aliquot. Scaling up for the full 250 $cm^3=8.08\times10^{-6}$ moles in the full 100 g sample of salt. Converting to mass (each iodate has one iodine atom), $8.08\times10^{-6}\times127=1.027\times10^{-3}$ g per 50 g of salt, which is 1027 μg per 50 g of salt or 2054 μg per 100 g of salt.

A7.

$$4NaI + O_2 + 2CO_2 \rightarrow 2Na_2CO_3 + 2I_2$$

a) Oxygen is the oxidizing agent, iodide is oxidized.
b) We could use starch indicator to see if iodine has been formed.
c) In IO_3^-, iodine is in a higher oxidation state and could not be oxidized further by oxygen in the air.

A8. To answer this question, we use $A = A_0\ e^{-(0.693/T_{1/2})}$. It is easier to deal with this equation if we take natural logs (ln):

$\ln A = \ln A_0 - 0.693t/T_{\frac{1}{2}}$.
4 weeks = 28 days, so $t = 28$. Therefore:
$\ln A = \ln(3.7 \times 10^{12}) - (0.693 \times 28)/8.04$.
$\ln A = 28.94 - 2.413 = 26.527$.
So, $A = e^{26.527} = 3.32 \times 10^{11}$ Bq after four weeks.

A9.

a) The rate equation is: Rate $= k[H_2O_2][I^-]$.
b) Thus, $k = \text{Rate}/[H_2O_2][I^-]$, making the units (mol dm^{-3} s^{-1})/(mol^2 dm^6); cancelling the units gives dm^3 mol^{-1} s^{-1}.
c) Step 1. $H_2O_2 + I^- \rightarrow H_2O + IO^-$.
Step 2. $IO^- + H_2O_2 \rightarrow H_2O + O_2 + I^-$.
The rate determining step is almost certainly step 1 as both species in the rate equation appear in the first step. Once the species have been formed, the second step is comparatively fast.

A10.

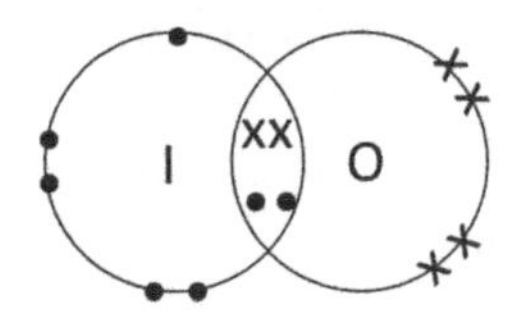

Iodine monoxide is a radical (it has an unpaired electron). Radicals are chemically reactive species.

CHAPTER 9

A1. (Ar) represents all electrons from $1s^2$ to $3p^6$

a)

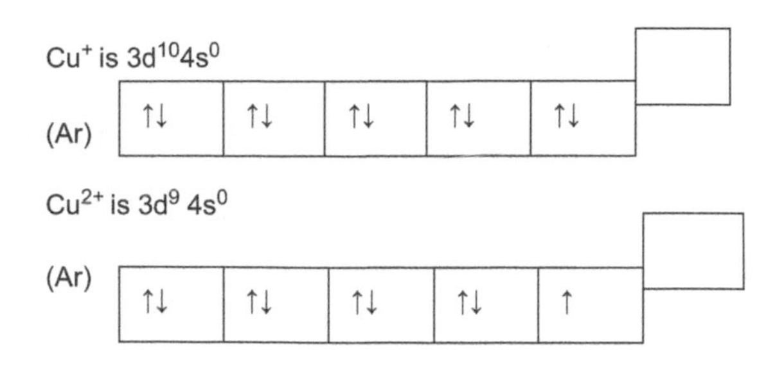

b) The colour of the Cu^{2+} ions in solution comes from the electronic transitions between degenerate 3d orbitals that form when ligands form dative bonds with the central Cu^{2+} ion. In Cu^{+} all of the 3d orbitals are full, and thus there are no electronic transitions between the degenerate (not of equal energy) levels that co-respond to absorbance in the visible spectrum. Aqueous Cu^{2+} ions transmit blue light because they absorb the complementary colour yellow.

A2. By introducing atoms of different sizes to the metal crystal, the regular layered structure is disrupted causing the different size metals to interlock in such a way as to reduce the ability of the layers to slide over each other.

A3.

a) $Cu_2CO_3(OH)_2 \rightarrow 2CuO + H_2O + CO_2$ = thermal decomposition. $2CuO + C \rightarrow 2Cu + CO_2$ = redox.

b) 500 kg of malachite $= 5\times10^5/221 = 2262.4$ moles of malachite, which will give an equal number of moles of Cu, which has a mass of 143.67 kg.

c) Atom economy = (mass of desired product/total mass of products)$\times$100 = (127/171)$\times$100 = 74.3%.

A4. $Fe^{2+}_{(aq)} + 2e^- \rightarrow Fe_{(s)}$ $E^{\theta} = -0.44$ eV
$Cu^{2+}_{(aq)} + 2e^- \rightarrow Cu_{(s)}$ $E^{\theta} = +0.34$ eV
The feasible reaction shown by the positive E^{θ}_{CELL} occurs when the top reaction is reversed:

$$Fe_{(s)} \rightarrow Fe^{2+}{}_{(aq)} + 2e^- \quad E^\theta = +0.44 \text{ eV}$$

$$Cu^{2+}{}_{(aq)} + 2e^- \rightarrow Cu_{(s)} \quad E^\theta = +0.34 \text{ eV}$$

$$Fe + Cu^{2+}{}_{(aq)} \rightarrow Cu + Fe^{2+}{}_{(aq)} \quad E^\theta = +0.78 \text{ eV}$$

A5.

a) Cu_5FeS_4 percentage mass Cu = (5×63.5)/((5×63.5) + 56 + (32×4))×100 = 63.3%.
b) From both equations, 4 moles of SO_2 are produced for every 2 moles of copper. 1000 kg of Cu = 1 000 000/63.5 = 15748 moles. Thus, 2×15748 moles = 31496 moles of SO_2, which occupies a volume of 31496×24 dm^3 = 755905 dm^3 or 756 m^3.
c) 200 Amps for 18 h = 200×18×60×60 = 1.296×10^7 Coulombs, which is 1.296×10^7/96,500 = 134.3 moles of electrons, which will deposit 134.3/2 = 67.15 moles of copper with a mass of 67.15×63.5 = 4264 g or 4.26 kg.
d) The mass collected would be less as there will be impurities that precipitate out in the electrolyte, such as silver.

A6.

a) $Zn_{(s)} + Cu^{2+}{}_{(aq)} \rightarrow Zn^{2+}{}_{(aq)} + Cu_{(s)}$.
b) 50 cm^3 of 1.00 M $CuSO_{4(aq)}$ is 0.05 moles of Cu^{2+} ions. We will need a minimum of 0.05×65 = 3.25 g of Zn to ensure that all of the copper sulphate reacts; thus, 3.5 g+ is excess.
c) Using $Q = mc\Delta T$, Q = 50×4.2×39.8 = 8358 J. This is for 0.05 moles of Cu^{2+} ions, so for 1 mole: 8358/0.05 = 167160 J, giving a ΔH_R of −167.2 kJ mol^{-1}.
d) The result represents about 60% of the data book value. Discrepancies will be due to heat lost to the surroundings and the calorimeter and non-standard conditions for reactants and products.

A7. Given $\Delta G = \Delta H - T\Delta S$, ΔS represents change in the disorder of the system. In this case reactants consist of 1 mole of solid + 1 mole of aqueous solution. The products are also 1 mole of solid + 1 mole of aqueous solution, and thus the change in entropy is likely to be small in comparison with reactions in which a) gases are produced or b) there is an increase in the numbers of moles of products.

A8.

a) $Cu^{2+}{}_{(aq)} + 2e^- \rightleftharpoons Cu_{(s)}$ $E^\theta = +0.34$ eV.
$Zn^{2+}{}_{(aq)} + 2e^- \rightleftharpoons Zn_{(s)}$ $E^\theta = -0.76$ eV.
The overall equation for the reaction is: $Zn_{(s)} + Cu^{2+}{}_{(aq)} \rightarrow Zn^{2+}{}_{(aq)} + Cu_{(s)}$, which gives an $E^\theta{}_{CELL}$ of $+1.10$ eV.
Using $\Delta G = -nFE^\theta$, we get $\Delta G = -2 \times 96\,500 \times 1.1 = 212\,300$ J or -212.3 kJ mol^{-1}.

b) The similarity between this value for ΔG and the data book value of -217 kJ mol^{-1} for ΔH confirms our expectation that the entropy change for this reaction is small.

A9.

a) Formal oxidation numbers shown below the reaction:

$$\underset{-0.5}{2O_2^-} + 2H^+ \rightarrow \underset{0}{O_2} + \underset{-1}{H_2O_2}$$

b) This would be classed as a disproportionation reaction.

c) The catalytic properties of many transition metals are based on their ability to "oscillate" between oxidation states.

A10. 78 mg = 0.078 g in 1850 cm^3. 0.078/63.5 = 0.00123 moles in 1850 cm^3, which is 6.64×10^{-4} moles per dm^3 or 664 μmol dm^{-3}.

CHAPTER 10

A1.

a) Compound 1 is fluoxetine, also known a Prozac, a member of a family of selective serotonin reuptake inhibitor (SSRI) medicines used to treat depression and anxiety.

Compound 2 is halothane, a compound used as a general anaesthetic.

b) 1-Bromo-1-chloro-2,2,2-trifluroethane.

c) Prozac: $C_{17}H_{18}NOF_3$, $M_R = 309$ g mol^{-1}, 18.4% F. Halothane: $C_2HBrClF_3$, $M_R = 197.5$ g mol^{-1}, 28.9% F.

d) The wavy line is attached to a carbon that has the potential to show optical activity.

e) The 3 : 4 : 1 ratio represents the relative proportions of the molecular ions with the different isotopes of Br and Cl. There is one way of obtaining the M^+ peak at 200, which is the ^{37}Cl isotope and the ^{81}Br isotope. The probability of getting this combination is based on the product of their isotopic abundance, which is (0.25×0.5). There are two possible ways of obtaining the M^+ 198 peak, (^{35}Cl and ^{81}Br) and (^{37}Cl and ^{79}Br). The cumulative probability is $(0.75 \times 0.5) + (0.25 \times 0.5)$. Finally, there is only one way of obtaining the M^+ 196 peak, which is (0.75×0.5). This is ^{35}Cl and ^{79}Br isotopes in the molecular ion.

f) The percentage abundance of ^{13}C is 1.1%. A molecule with two carbons would, therefore, have a 2.2% chance of having a ^{13}C atom present. The possibility of both being ^{13}C is, however, very small (0.011×0.011), and is thus unlikely to add to any of the molecular ion peaks.

A2.

a) $CaF_2 = 2.3 \times 10^{-5}$; $CaCl_2 = 5.36 \times 10^{-1}$.
K_{sp} for $CaF_2 = [Ca^{2+}][F^-]^2 = 2.31 \times 10^{-5}$. $\sqrt[3]{2.31} \times 10^{-5} =$ 0.0285 moles per 100 cm^3 of water, which is 2.223 g.
K_{sp} for $CaCl_2 = [Ca^{2+}][Cl^-]^2 = 5.36 \times 10^{-1}$. $\sqrt[3]{5.36} \times 10^{-1} =$ 0.812 moles per 100 cm^3 of water, which is 90.2 g.

b) Due to the common ion effect, the high concentration of the Ca^{2+} ions in solution from $CaCl_2$ will reduce the solubility of CaF_2.

Think about how a high concentration of Ca^{2+} ions affects the following equilibrium:

$$CaF_{2(s)} \rightleftharpoons Ca^{2+}{}_{(aq)} + 2F^-{}_{(aq)}$$

A3. No set answers from me on this one, it's up to you! What research I have looked at suggests that water companies tend not to bother these days.

A4.

a)

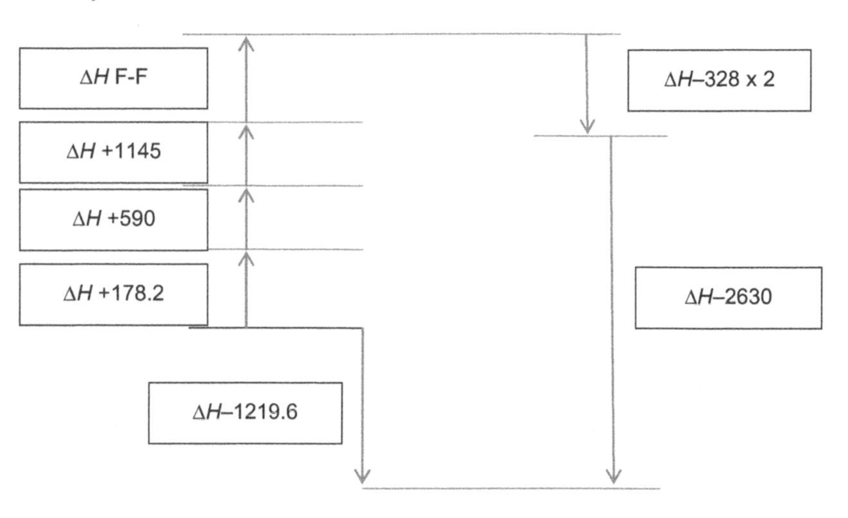

Thus, (1219.6) + (178.2) + (590) + (1145) + F–F = (328×2) + (2630).
F–F = 3286 − 3128.8 = 157.2 kJ mol^{-1}.

b) Halogen bond enthalpies are as follows:

Halogen, X–X	*Bond enthalpy/kJ mol^{-1}*
F–F	157.2
Cl–Cl	243.4
Br–Br	192.9
I–I	151.2

The F–F bond does not fit the trend. This is explained, in part, by the forces of repulsion between the two F nuclei, which is large due to the short bond F–F bond length.

A5.

a) The IUPAC names and oxidation states are:
a) $CaSO_3$: calcium sulfate IV (S = +4).
b) K_2MnO_4: potassium manganite VI (Mn = +6).
c) $(NH_4)_2MoO_4$: ammonium molybdate VII (Mo = +7).
d) Cl_2O_7 : chlorine VII oxide (Cl = +7).

e) $\underline{P}_4O_{10}$: phosphorus V oxide (P = +5).
f) $\underline{C}Cl_4$: tetrachloromethane (C = +4).
g) $\underline{F}_2O$: fluorine I oxide (F = −1).

b)

$$\underset{0}{2F_2} + \underset{+1\ -2}{2H_2O} \rightarrow \underset{+1\ -1}{4HF} + \underset{0}{O_2}$$

Yes, oxygen is oxidized!

A6.

a)

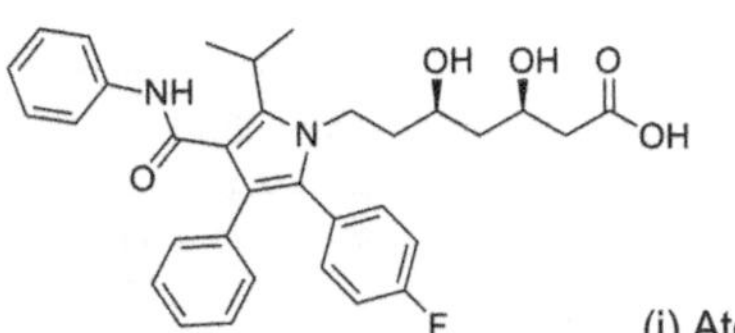

(i) Atorvastatin

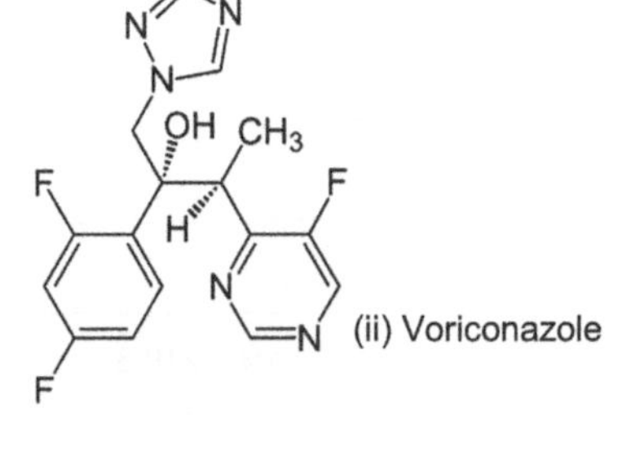

(ii) Voriconazole

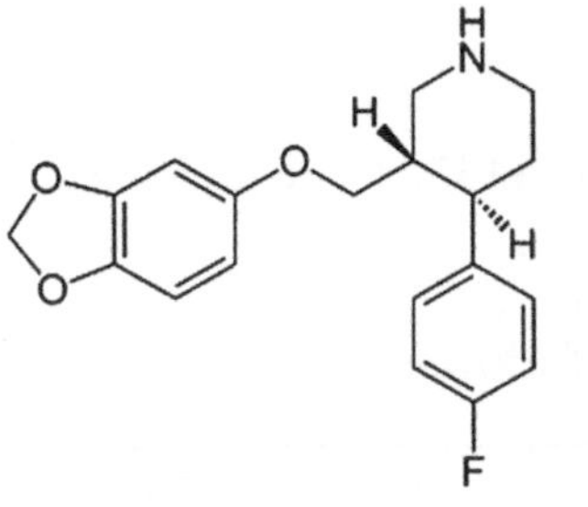

(iii) Paroxetine

b) In all three cases one or more F atoms is directly attached to an aromatic group.

A7.

a)

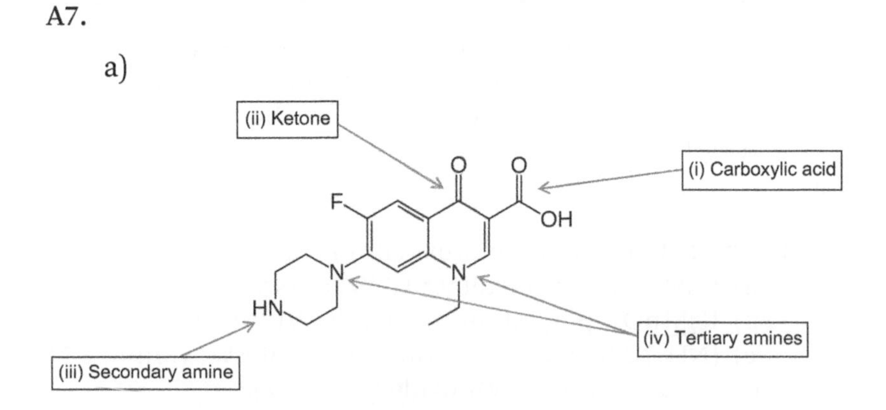

b) Eighteen H atoms and nine different proton environments.

A8.

a)

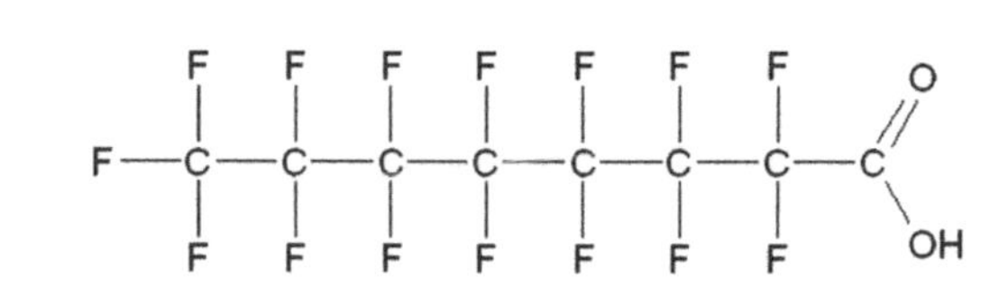

b) 400 ng is 4.0×10^{-7} g. This is $4.0\times10^{-7}/414=9.66\times10^{-10}$ moles$\times3.5=3.38\times10^{-9}$ moles.

A9.

a) The electron withdrawing effect of the F atom has the effect of increasing the polarity of the carboxylic acid O-H group and stabilizes the carboxylate ion formed.
b) pK_a for ethanoic acid $=4.76$. Thus, $K_a=1.74\times10^{-5}$ mol dm^{-3}. For a weak acid at 0.01 M dilution, we can assume that $\alpha^2/(\text{concentration}-\alpha)$ is $\approx\alpha^2/\text{concentration}$. Thus, $1.74\times10^{-5}=\alpha^2/0.01$, so $\alpha=\sqrt{1.74\times10^{-7}}=4.17\times10^{-4}$. Thus, $\text{pH}=-\log_{10}(4.17\times10^{-4})=3.38$.
c) and d) For fluoroethanoic acid, the degree of dissociation is sufficiently great such that we cannot make the $\alpha^2/(\text{concentration}-\alpha)\approx\alpha^2/\text{concentration}$ approximation. Consequently, we have to solve:
$pK_a=2.66$, so $K_a=2.19\times10^{-3}=\alpha^2/(0.01-\alpha)$ and solve for α.

A10.

a) All bond angles are 90°.

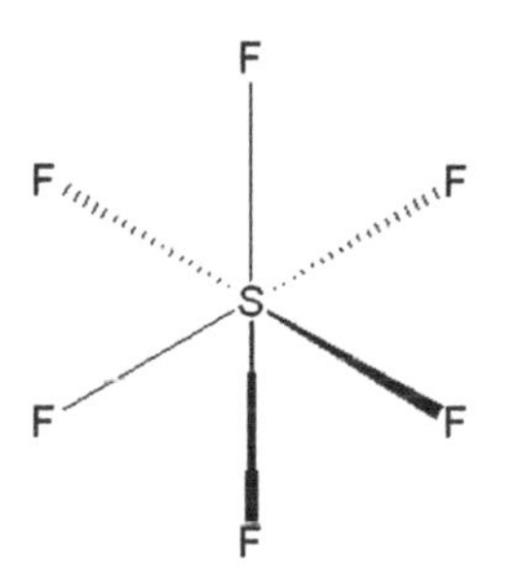

b) Sulfur has an oxidation number of +6.

A11.

a) Phosphorus oxidation number = +5.

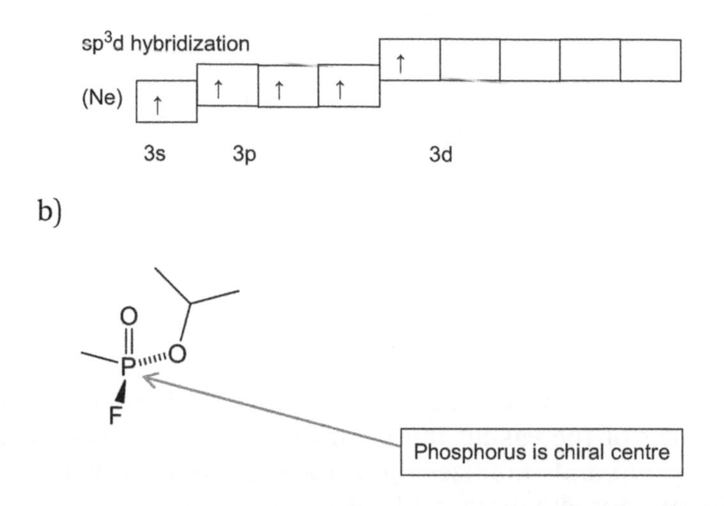

b)

A12. There are through-space effects (as opposed to through-bond effects) that can affect the electron shielding of an atom's nucleus, and thus its resonance frequency. In a folded protein the close proximity of the F atom to other groups may cause the specific resonance frequency to be shifted. This can be used as in indicator of whether a protein is folded or not.

CHAPTER 11

A1.

a) Amphoteric substances are substances that can react with both acids and bases to form salts.
b) i) $Al_2O_3 + 6HCl \rightarrow 2AlCl_3 + 3H_2O$.
ii) $Al_2O_3 + 2NaOH + 3H_2O \rightarrow 2NaAl(OH)_4$.
c) $Al = 1s^2 2s^2 2p^6 3s^2 3p^1$; $Al^{3+} = 1s^2 2s^2 2p^6$. It is isoelectronic (has the same electronic configuration) with Ne.
d)

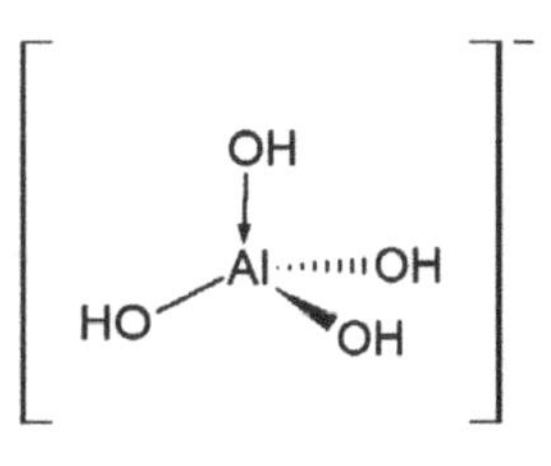

The aluminium atom is sp^3 hybidized. One dative bond is formed with OH^-.
e) $Al + NaOH + 3H_2O \rightarrow NaAl(OH)_4 + H_2$; 1 : 1 reaction ratio. 50 kg = 50 000/27 = 1851.85 moles of Al. This requires the same number of moles of NaOH. $C = n/V$, so $6.0 = 1851.85/V$, which gives $V = 308.64$ dm^3 of 6.0 M NaOH required.

A2.

a) One kg of Al is 1000/27 = 37.037 moles of Al. This requires 3×37.037 moles of electrons = 111.1 moles of electrons. One mole of electrons is 96 500 C, so $111.1 \times 96\,500 = 1.072 \times 10^7$ C. $J = C \times V$, so at 6 V, the total number of joules needed $= 6 \times 1.072 \times 10^7 = 6.43 \times 10^7$ J or 6.43×10^4 kJ. One kW h = 3600 kJ, so $6.43 \times 10^4 / 3600 = 17.861$ kW h.
b) At 15p per kW h, electricity for 1 kg of Al $= 17.861 \times 15 =$ £2.68 to the nearest p.
c) The overall equation is: $2Al_2O_3 \rightarrow 4Al + 3O_2$. One kg of Al = 37.037 moles of Al. The equation (eqn 11.2) shows that for every 4 moles of Al 3 moles of Al formed. So $\frac{3}{4} \times 37.037 = 27.78$ moles of O_2. This would oxidize an equal number of moles of carbon, which is $27.78 \times 12 = 333.3$ g.

A3.

a) An electrochemical cell is set up in which the aluminium acts as a sacrificial anode and reduces the oxidized silver back to silver metal:

$$Al^{3+}{}_{(aq)} + 3e^{-} \rightarrow Al_{(s)} \quad -1.66.$$

$$Ag^{+}{}_{(aq)} + e^{-} \rightarrow Ag_{(s)} \quad +0.80.$$

b) Overall reaction: $Al_{(s)} + 3Ag^{+}{}_{(aq)} \rightarrow 3Ag_{(s)} + Al^{3+}{}_{(aq)}$ $E_{CELL} = +2.46$ eV.

A4. $KAl(SO_4)_2 \cdot 12H_2O_{(s)}$; $M_R = 474$ g mol^{-1} . 237/474 = 0.5 moles. The solution is thus 0.5 M with respect to $[Al(H_2O)_6]^{3+}$ ions. Thus, taking this aqua ion as a weak acid:
$\alpha^2/0.5 = 1\times10^{-5}$, so $\alpha = \sqrt{5\times10^{-6}} = 2.23\times10^{-3}$. So pH = 2.65.

A5.

a) $AlCl_3$ is trigonal planar with 120° bond angle:

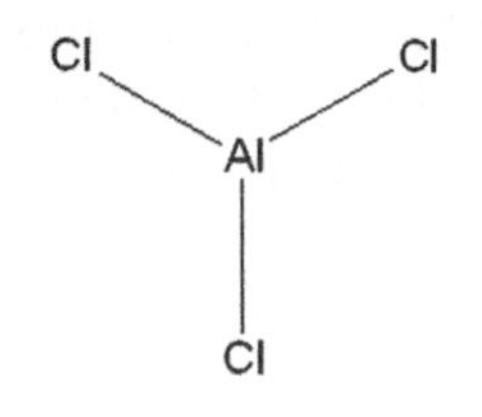

b) A Lewis acid is an electron pair acceptor.

c) The Al_2Cl_6 dimer forms due to a lone pair from one of the Cl atoms, forming a dative covalent bond with a second Al in an adjacent $AlCl_3$ molecule. This is reciprocated so that two dative bonds are formed as shown below. All bond angles are 109.5°.

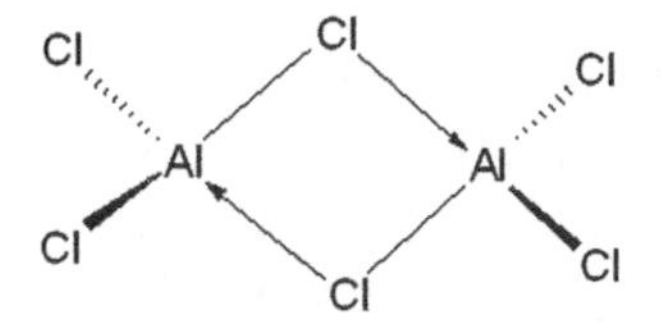

A6.

a) $2Al_{(s)} + Fe_2O_{3(s)} \rightarrow Al_2O_{3(s)} + 2Fe_{(l)}$.
$\Delta H^{\theta}{}_{R} = -1675.7 - (-824.2) = -851.5$ kJ mol^{-1}.

b) 2.5 kg of Fe is 2500/56 = 44.643 moles. We need 1/2 as many moles of Fe_2O_3 so $1/2 \times 44.643 \times 160 = 3571.4$ g or 3.6 kg Fe_2O_3 and $44.643 \times 27 = 1205.4 = 1.21$ kg.

CHAPTER 12

A1.

a) Hydrogen forms diatomic (H_2) molecules. As a solid, the forces of attraction that hold the molecules together are dispersion forces, which are formed due to the induced dipole between the H_2 molecules. As there are only two electrons per molecule, these dispersion forces are very weak, and thus are only significant at very low temperatures when the molecules have very little kinetic energy. Lithium forms a metallic bond in which lithium ions are attracted by a sea of delocalized electrons, which are formed when the outer $2s^1$ electron is delocalized throughout the metal crystal.

b) The suggestion here is that, under extreme pressure, the H_2 molecules no longer form covalent bonds but form something similar to a metallic bond, like lithium.

A2.

a) Using $E = mc^2$, $E = (1.5 \times 10^{13}) \times (3.0 \times 10^8)^2 = 1.35 \times 10^{30}$ J or 1 350 000 YJ in 1 h.

b) Power is measured in J s^{-1} or W. 1 350 000/3600 = 375 YW (yottawatts).

c) kg$\times$(m $s^{-1})^2$. But 1 kg m^2 s^{-2} = 1 N (Newton). Thus, kg$\times$(m $s^{-1})^2$ = Nm, which is 1 J.

A3.

a) $C_xH_Y + xH_2O \rightarrow xCO + (y/2 + x)H_2$.

b) $C_4H_{10} + 4H_2O \rightarrow 4CO + 9H_2$.

A4.

a)

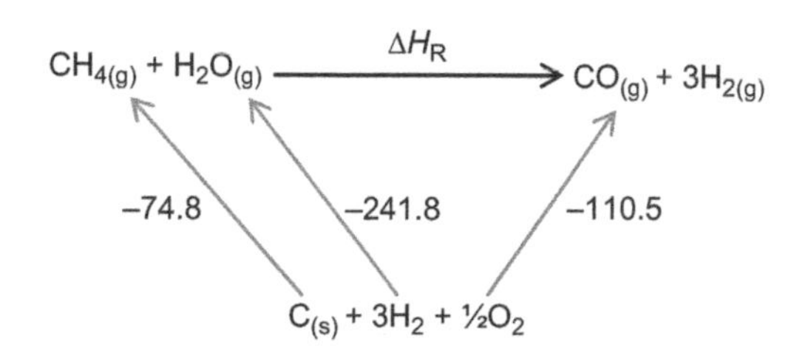

$\Delta H_R = (+74.8) + (+241.8) + (-110.5) = +206.1$ kJ.

b) Because the forward reaction is endothermic, a high temperature favours the forward reaction, and thus a higher yield of hydrogen.
c) However, the second step is an exothermic reaction, and thus a higher temperature will favour the reverse reaction.
d) In both reactions steam is a reactant. If steam is in a large excess, then the forward reaction will be favoured in both cases.

A5.

a) $K_p = p(H_2) \times p(CO_2)/p(CO) \times p(H_2O)$. There are no units because the combined stoichiometric values are equal on both sides of the equation.
b) Taking $\ln K = -\Delta H/RT + c$ and using ratios, the c constant cancels. Putting our known values in the equation and cancelling: $\ln(8.13)/\ln(\alpha) = -700/1100$ where $\alpha = K_p$ at 1100 K. Solving for $\ln(\alpha) = -1100/700 \times \ln(8.13) =$ so $\ln(\alpha) = -3.29$ thus $\alpha = e^{-3.29}$ thus K_p at 1100 K = 0.0373.
c) The smaller value for the K_p at 1100 K tells us that there are less products than at 700 K, which means that the equilibrium has shifted to the left. This is what we would expect from a qualitative prediction as an increase in temperature will favour the endothermic reaction, which is the reverse reaction in this case.
d) A manufacturer must consider the rate of reaction and the energy costs of maintaining reaction conditions.

A6.

a) Oxygen is sp^3 hybridized with two bonding pairs and two lone pairs in its valence shell. All four pairs adopt a tetrahedral shape but the lone pairs exert a greater repulsive effect than the bonding pairs and distort the HOH bond angle to 104.5°.
b) As a result of its "V" shape and the difference in electronegativity between oxygen and hydrogen, the water molecule has a permanent dipole. Water can donate two electron pairs to form hydrogen bonds and accept two electron pairs, forming another two hydrogen bonds and making a total of four hydrogen bonds per molecule. The resultant intermolecular forces are strong enough to attract water molecules together right up to 100 °C at 1 atm. pressure. Sulfur dioxide may have a larger molar mass but

the intermolecular forces are weaker because the difference in electronegativity between S and O is considerably smaller than for H and O.

A7. The fact that water expands when it freezes is due to the regular hydrogen bonding that forms when the liquid turns into a solid. This means that frozen water floats on liquid water and traps any residual heat in. During periods in which the Earth was frozen (the so called snow ball Earth), the icy crust would have protected any life-forms from freezing.

A8.

a) At 60 °C the $K_w = 5.6\times10^{-14}$ mol^2 dm^{-6} for water $=$ $[H^+][OH^-]$. So $[H^+] = \sqrt{5.6\times10^{-14}} = 2.37\times10^{-7}$, which makes $\log_{10}(2.37\times10^{-7}) = 6.63$.

b) If the pH decreases, then the degree of ionization must be increasing. This suggests that the forward reaction must be endothermic because the equilibrium in responding to the increase in temperature by trying to absorb the extra temperature, thus favouring the forward reaction.

A9.

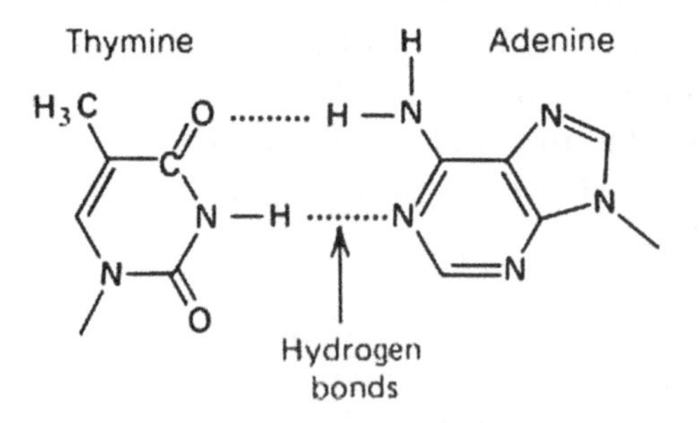

A10. The skeletal formulae of the two organic structures are shown below:

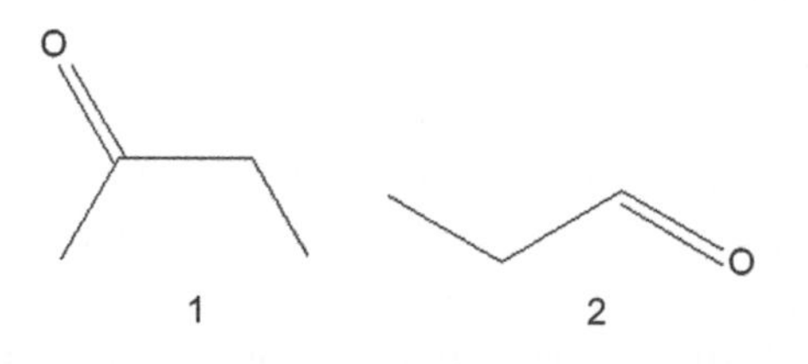

a) The environment of the proton is determined by the degree of shielding that it experiences due to other atoms' influence on the hydrogen atoms valence electrons. This influences can act through covalent bonds or act through space.

b) (i) 1 = Butanone and 2 = propanal.

(ii) Butanone would have three difference proton environments with relative areas under the peaks of 3 : 2 : 3 (from left to right as drawn) The methyl protons on the left of the molecule, as drawn, would give rise to a single peak (no splitting) with a relative area of 3. The CH_2 protons would give rise to a quartet (1 : 3 : 3 : 1) with a relative area of 2 and the final methyl protons would give rise to a triplet (1 : 2 : 1) with a relative area of 3.

Propanal would also give three proton environments but at different chemical shifts due to the different proton environments. The aldehyde proton would give rise to a triplet and the terminal CH_3 protons would give rise to a triplet. The central CH_2 protons are more complex as they split both with the methyl protons (to give a quartet) and then these are themselves split due to the aldehyde proton. Using high resolution NMR, you could expect to see a quartet with each peak split into a doublet.

CHAPTER 13

A1.

a)

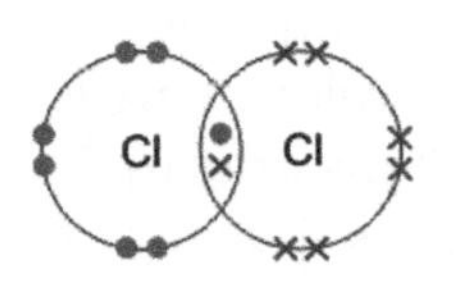

b) Dispersion forces, or temporary induced dipole interactions, occur due to the constant motion of electrons in the Cl_2 molecule. At any one moment in time, the electron distribution will be unequal and result in a temporary dipole. This, in turn, will induce a dipole in an adjacent molecule and result in intermolecular forms of attraction.

c) For molecules without permanent dipoles, it is the constant flux of the electrons that causes the dispersion forces. In general, the greater the number of electrons, the stronger the dispersion forces. Consequently, there is trend in which the boiling point increases as atomic number increases from F to At.

A2.

a)

$$MnO_2 + 4H^+ + 2e^- \rightleftharpoons Mn^{2+} + 2H_2O \quad E^\theta = +1.23 \text{ eV}$$

$$2Cl^- \rightleftharpoons Cl_2 + 2e^- \quad E^\theta = -1.36 \text{ eV}$$

$$MnO_2 + 4H^+ + 2Cl^- \rightleftharpoons Mn^{2+} + 2H_2O + Cl_2 \quad E^\theta_{CELL} = -0.13 \text{ eV}$$

The $E^\theta{}_{CELL}$ is negative (–0.13 eV), which suggests that the reaction is not feasible under standard conditions.

b) Standard conditions are 298 K, 1 atm. pressure and all solutions are 1.0 M with respect to aqueous ionic species.

c) Using $E = E^\theta - \frac{RT}{nF}\ln Q$, where Q = [reduced species]/[oxidized species] we can calculate the effect of changing the temperature and chloride ion concentration on the chlorine/chloride ion half cell:

$$E = (+1.36) - (8.314 \times 353)/(2 \times 96\,500) \times \ln(6)^2/1.$$

$$E = (+1.36) - 0.0152 \times 3.5835 = +1.36 - 0.054 = +1.306.$$

And for the MnO_2/Mn^{2+} half cell:

$$E = (+1.23) - (8.314 \times 353)/(2 \times 96\,500) \times 1/\ln(6)^4.$$

$$E = (+1.23) - (0.0152 \times (-7.167) = +1.23 + 0.109 = +1.339.$$

d) Under these conditions, the new E_{CELL} is now $1.339 - 1.306 = +0.033$ and is now feasible.

A3.

a)

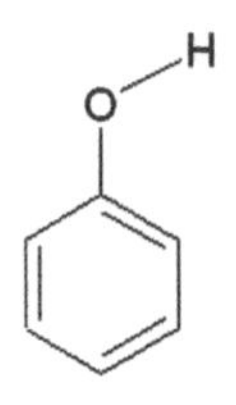

b)

Three possible resonance structures of the phenoxide ion are shown. The ability of the ion to spread the negative charge over a number of possible structures stabilizes the ion and makes the phenoxide ion a weaker conjugate base.

c) $K_a = [C_6H_5O^-][H^+]/[C_6H_5OH] = 1.28 \times 10^{-10}$.
$\alpha = \sqrt{(1.28 \times 10^{-10})} \times 0.002 = \sqrt{2.56 \times 10^{-13}} = 5.06 \times 10^{-7}$.
$pH = 6.30$.

d)

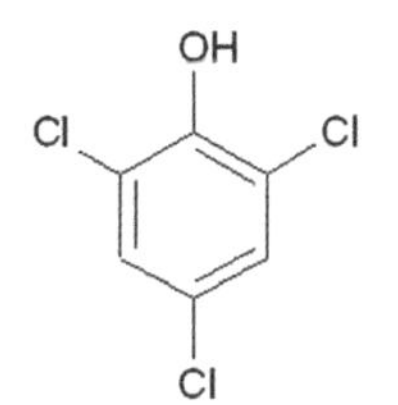

e) As the OH group activates the benzene ring no Freidel–Crafts catalyst is required to carry out an electrophillic substitution and produce the chlorinated phenol. However, as the chlorine has an electron withdrawing effect second and third substitutions become more difficult. Warming the phenol and reacting with chlorine gas should be enough to produce Trichlorophenol although there will inevitably be mono and dichlorophenols in the mix.

f) There are two proton environments. The two aromatic protons are equivalent and would give rise to a peak twice the height of the OH peak, which would have a greater chemical shift due to the electron-withdrawing effect of the oxygen atom.

g) The combined effect of the three chlorine atoms at positions 2, 4 and 6 draws the electrons toward the aromatic ring so that the OH is more easily accepted by a water molecule. Consequently, TCP has a larger K_a value.

A4.

a) Elemental Cl_2 has a 0 oxidation number. As a result of the reaction with water, it is both oxidized (HOCl = +1) and reduced (HCl = –1). Simultaneous reduction and oxidation of the same species is disproportionation.

b) $HOCl \rightleftharpoons H^+ + OCl^-$, $K_a = [H^+][OCl^-]/[HOCl]$; *i.e.*, $pK_a = 7.4$ and the $K_a = 1 \times 10^{-7.4} = 3.98 \times 10^{-8}$.

c) Using the relationship: $pH = pK_a - \ln(HOCl)/\ln(ClO^-)$, it becomes clear that, at half neutralization of HOCl, the $pH = pK_a$, which is 7.4 at standard temperature and pressure.

d) HOCl is a neutral species and is more likely to be absorbed by the cell membrane, whereas the ClO^- ion is negative, and thus is likely to be repelled by the negatively charged cell membrane.

A5. The following data is used to answer this question:

$2H^+ + S + 2e^- \rightarrow H_2S$	+0.14
$Cl_2 + 2e^- \rightarrow 2Cl^-$	+1.36

The overall feasible reaction is: $Cl_2 + H_2S \rightarrow 2Cl^- + 2H^+ + S + 1.22$.
The oxidation of H_2S to S removes the rotten egg smell of H_2S.

A6.

a) $k = \text{rate}/[ClO^-]^2$, so mol dm^{-3} s^{-1}/mol^2 dm^{-6} = dm^3 mol^{-1} s^{-1}.
b) Probably step 1 as it involves two ClO^- ions.
c) Taking natural logs, $\ln k = \ln A - E/RT$.
$\ln(11.4) = \ln(2.1 \times 10^{12}) - E/(8.314 \times 333)$.
$E/(8.314 \times 333) = 28.37 - 2.434$, so $E = 71.8$ kJ mol^{-1}.

A7.

a) Empirical formula 24.24/12 : 4.04/1 : 71.71/35 = 2.02 : 4.04 : 2.04 so CH_2Cl.
b) Molecular ion peaks at ratios of 9 : 6 : 1 suggests two chlorine atoms in molecule representing statistical probabilities of ^{35}Cl and ^{37}Cl isotopes.
c) So molecular formula $C_2H_4Cl_2$.
d) If the rate determining step requires a collision between two ClO^- ions, then dilution will reduce the frequency of collisions.

A8.

a)

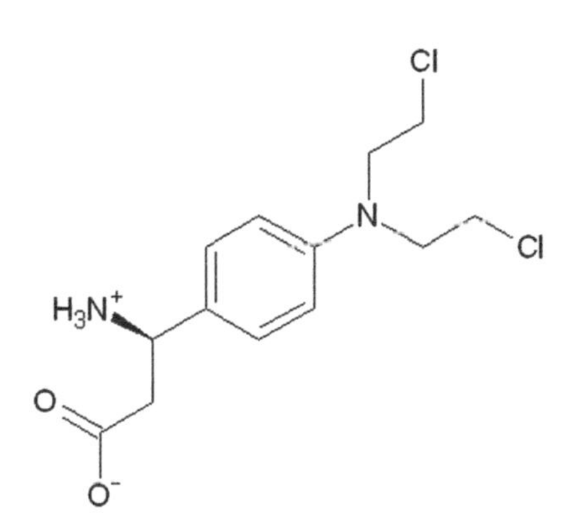

The charged groups make more energetically favourable electrostatic interactions with water molecules, thus improving solubility.

b)

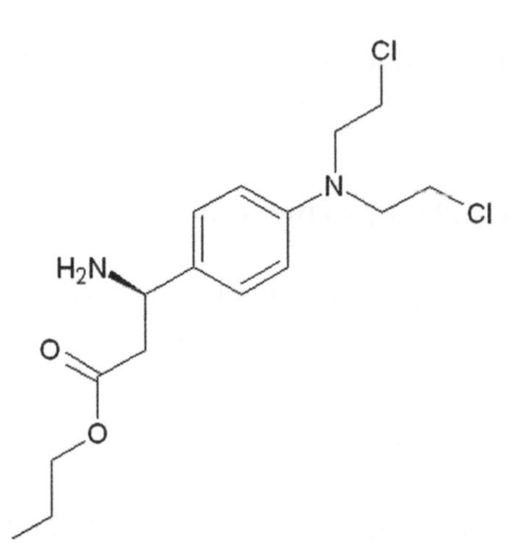

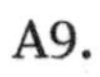
A9.

a)

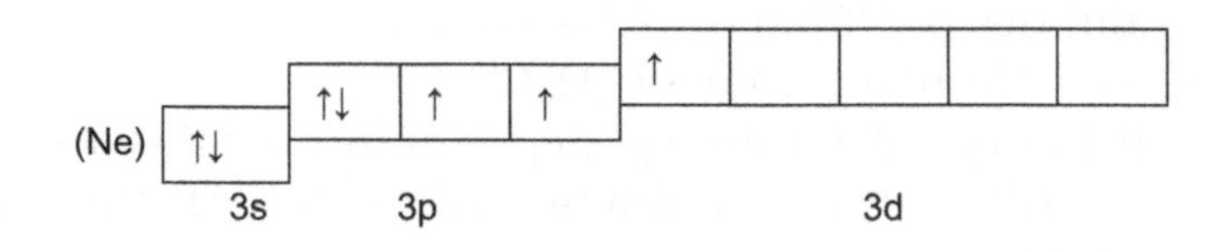

b)

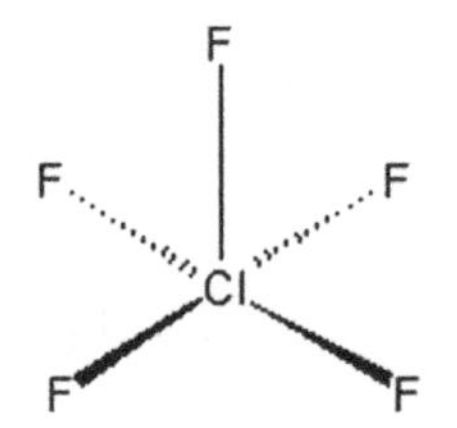

All bond angles are 90° with a lone pair below the square plane of the molecule.

A10.

a) 60% of 5 $dm^3 = 3$ dm^3. At a concentration of up to 116 mmol dm^{-3}, this is $0.116 \times 35.5 = 4.12$ g.
b) $4.12/95 \times 100 = 4.33\%$.

A11.

a)

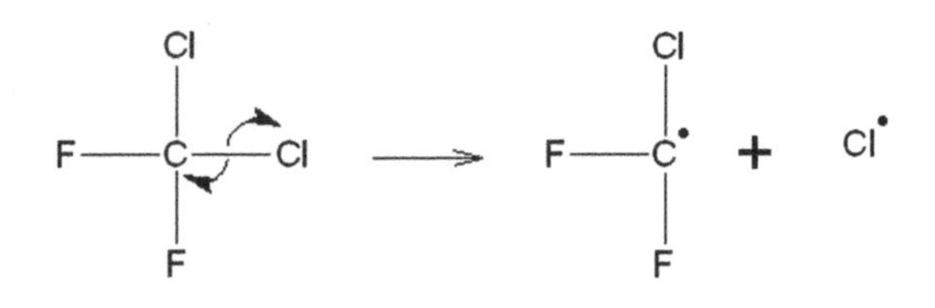

b) $Cl^{\bullet} + O_3 \rightarrow ClO^{\bullet} + O_2$.
$ClO^{\bullet} + O \rightarrow O_2 + Cl^{\bullet}$.

c) Ultraviolet light has insufficient energy to break the more stable C–F bond.

d) Both ClO and NO_2 are radicals and thus, by combining two radicals, cancel each other out.

CHAPTER 14

A1.

a) $Zn = 1s^2 2s^2 2p^6 3s^2 3p^6 3d^{10} 4s^2$; $Zn^{2+} = 1s^2 2s^2 2p^6 3s^2 3p^6 3d^{10}$.
b) A transition metal is defined as a metal that can form at least one ion with an incomplete set of d orbitals.
c) 

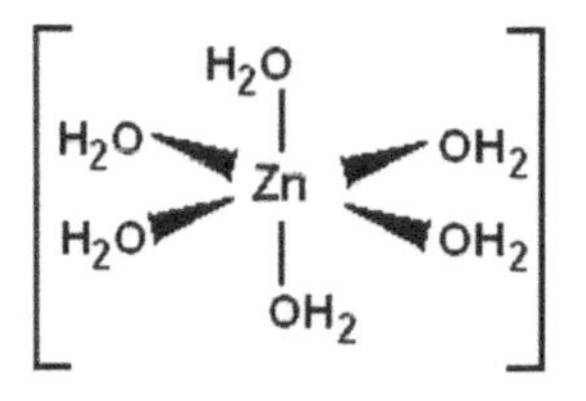

The ion is octahedral. As the 3d subshell is full ($3d^{10}$), there are no possible electronic transitions between degenerate 3d orbitals, and thus no visible frequencies of light absorbed.

A2. $(48.75/100)\times 64 + (27.9/100)\times 66 + (4.2/1000\times 67 + (19.15/100)\times 68 = 65.3$.

A3.

a) $ZnS_{(s)} + O_{2(g)} \rightarrow ZnO_{(s)} + SO_{2(g)}$.
b) 88% of 10 tonnes = 8.8 tonnes of ZnS. Mass ratio $ZnS/SO_2 = 97/64$. So, 97/64 = 8.8/mass of SO_2 = 5.81 tonnes.
c) $ZnO_{(s)} + H_2SO_{4(aq)} \rightarrow ZnSO_{4(aq)} + H_2O_{(l)}$.
d)

$$Zn_{(s)} \rightarrow Zn^{2+}{}_{(aq)} + 2e^- \quad +0.76$$

$$Cu^{2+}{}_{(aq)} + 2e^- \rightarrow Cu_{(s)} \quad +0.34$$

$$\text{So, } Cu^{2+}{}_{(aq)} + Zn_{(s)} \rightarrow Zn^{2+}{}_{(aq)} + Cu_{(s)} \quad +1.10$$

So, the reduction of copper ions by zinc is feasible.

A4.

a) 1.10 V (see A3c).
b) The electrolyte was sulfuric acid, and thus standard electrolytes were not used. There would probably be internal resistance in the cell as current is drawn.
c) The half cell equations show that $H^+{}_{(aq)}$ ions would be reduced before copper ions are deposited, and thus H_2 would build up at the cathode.

A5.

a) i) Cathode: $Ca^{2+} + 2e^- \rightarrow Ca_{(l)}$ and anode: $2O^{2-} \rightarrow O_2 + 4e^-$; ii) cathode: $K^+ + e^- \rightarrow K_{(l)}$ and anode: $4OH^- \rightarrow O_2 + 2H_2O + 4e^-$.
b) For electrolysis to occur, ions must be able to move. This can only happen with ionic compounds if they are molten.

A6.

a) $Cu^{2+} + 2e^- \rightarrow Cu_{(s)}$.
b) Using the data in Table 14.2, anything above 2.07 V would cause dissolved oxygen to react with the water to generate ozone and gradually decompose the water.

A7.

a) 298 K, 1 atm. pressure and all solutions 1.0 M with respect to all ions.
b) $Q = [Zn^{2+}_{(aq)}]/[Cu^{2+}_{(aq)}]$.
c) $E = E^\theta - RT/nF \ln Q$. If $\ln Q = 1$, then $E = E^\theta$.
d) $E = E^\theta - RT/nF \ln[1/0.1]$, thus $E = 1.10 - (8.314 \times 298)/(2 \times 96\,500) \ln 10 = 1.10 - (0.0128 \times 2.303) = 1.071$ eV.
e) 1. Decrease the concentration of the Zn^{2+} ions. $Zn^{2+}_{(aq)} + 2e^- \rightarrow Zn_{(s)} = -0.76$. A lower Zn^{2+} ion concentration would shift the equilibrium from right to left encouraging the oxidation of Zn.
2. Increase the concentration of the Cu^{2+} ions. $Cu^{2+}_{(aq)} + 2e^- \rightarrow Cu_{(s)} = +0.34$. Increasing the Cu^{2+} ion concentration would shift the half cell equilibrium to right, increasing the E^θ value.

A8.

a) The reaction involves 1 mole of solid metal and 1 mole of aqueous divalent metal ions reacting to form one mole of metal and one mole of aqueous divalent ions. Thus, changes in degrees of disorder are likely to be small.
b) As the forward reaction is exothermic and the increase in temperature is likely to favour the reverse reaction (*i.e.*, the endothermic process).
c) When aqueous reactants and products are both $\ln[Zn^{2+}]/[Cu^{2+}] = 0$ and so $E = E^\theta$ but, if the reaction moves to the left hand side, then $Q < 1$, which means $\ln[Q]$ is negative, and thus E increases with increasing T.

A9.

$$Zn + 2OH^{-}_{(aq)} \rightarrow ZnO_{(s)} + H_2O_{(l)} + 2e^{-} \quad E^{\theta} = +1.28\ eV$$

$$2MnO_{2(s)} + H_2O_{(l)} + 2e^{-} \rightarrow Mn_2O_{3(s)} + 2OH^{-}_{(aq)} \quad E^{\theta} = +0.15\ eV$$

$$2MnO_{2(s)} + Zn \rightarrow Mn_2O_{3(s)} + ZnO_{(s)} \quad E^{\theta}_{CELL} = +1.43\ eV$$

A10.

$Zn^{2+}_{(aq)} + 2e^{-} \rightarrow Zn_{(s)}$	$+0.76$
$Fe^{2+}_{(aq)} + 2e^{-} \rightarrow Fe_{(s)}$	-0.44
$Fe^{2+}_{(aq)} + Zn_{(s)} \rightarrow Fe_{(s)} + Zn^{2+}_{(aq)}$	**$E_{CELL} = +0.32$**

Zinc acts as a sacrificial anode. By oxidizing, it provides a source of electrons that keeps the iron from oxidizing.

A11.

a) Unlike copper, zinc adopts only one oxidation state ($+2$), and thus cannot switch between oxidation states like iron or copper can. This restricts its versatility as a catalytic centre.

b) 0.2 mM is 2×10^{-4} mol dm^3. Thus $(2\times10^{-4})\times(1.8\times10^{-15}) = 3.6\times10^{-19}$ moles per cell. Thus, $3.6\times10^{-19}\times6.022\times10^{23} = 216\ 792$ Zn^{2+} ions.

CHAPTER 15

A1.

a) The vapour pressure of a substance is the upward pressure exerted by the liquid moving from the liquid state to the vapour state when the system is at equilibrium.
b) Once the vapour pressure equals atmospheric pressure, the substance boils. Mercury would boil under these conditions.

A2. Percentage Hg in pure HgS $=(201/233)\times100=86.3\%$. Using ratios: $78 : 86.3 = x/100 = 90.4\%$ purity.

A3.

a) The reaction becomes feasible when $\Delta H - T\Delta S = 0$.
$\Delta H = 90.8/2 = 45.54$ kJ mol^{-1} of HgO decomposed.
$\Delta S = (76+102.5/2) - 70.3 = +56.95$ J K^{-1} mol^{-1}.
$T = \Delta H/\Delta S$, so $T = 45\,540/56.95 = 799.6$, *i.e.*, about 800 K or 527 °C.
b) At 350 °C, even though ΔG is negative, the value of the equilibrium constant is such that some of the mercury will be oxidized.
c) Even though, at room temperature, the equilibrium constant indicates that the reaction is feasible, the kinetics of the reaction are such that the rate of reaction at this temperature is almost 0. The reaction is thus said to be thermodynamically unstable but kinetically stable.

A4.

a) $2Cl^- \rightarrow Cl_2 + 2e^-$.
b) $Hg^{2+} + 2e + Na \rightarrow HgNa$.
c) $2HgNa + 2H_2O \rightarrow 2NaOH + H_2 + Hg$.

A5.

a) $HgO + Zn \rightarrow Hg + ZnO$ $E_{CELL} = 1.358$ eV.
b) $\Delta G = -2\times96\,500\times1.358/1000 = 262.1$ kJ mol^{-1}.
Using the relationship: $\Delta G = -nFE^0$, calculate the free energy change for this reaction.
c) $\Delta H_R = (-348.3) - (-90.8) = -258$ kJ mol^{-1}.
d) The two values for ΔG and ΔH are very close because the entropy change for this reaction is small. This is to be expected because the number of moles of reactants and

products is the same and there are no gases produced or consumed.

A6.

a) $Hg_2^{2+} \rightarrow Hg^{2+} + Hg_{(l)}$ $E^{\theta}{}_{CELL} = -0.131$.
The reaction is not feasible under standard conditions.

b) $\ln K = nFE^{\theta}/RT$, so $\ln K = 1 \times 96\,500 \times 0.131/8.314 \times 298 = 12641.5/2477.6 = 5.1$. $K = e^{-5.1} = 6.1 \times 10^{-3}$.
$K = [Hg^{2+}]/[Hg_2^{2+}] = 1 : 6.1 \times 10^{-3}$ or a ratio of approximately 1 : 164.

CHAPTER 16

A1. The states refer to solid, liquid and gas. However, it is possible to have two liquids that are physically separated due to their physiochemical properties. Oil and water, for example, are both liquids but there is a physical boundary that keeps the two from forming a homogenous mixture. They represent two distinct phases.

A2.

a) M_R $KMnO_4 = (39 + 55 + (16 \times 4)) = 158$. $1.58/158 = 0.01$ moles. Concentration $= 0.01/0.5 = 0.02$ mol dm^{-3}.

b) The top equation is multiplied by 2 and the second equation is multiplied by 5. The electrons then cancel out in the overall equation.

$$MnO_4{}^-{}_{(aq)} + 8H^+{}_{(aq)} + 5e^- \rightarrow Mn^{2+}{}_{(aq)} + 4H_2O_{(l)} \quad E^\theta = +1.51 \text{ eV}$$

$$H_2C_2O_{4(aq)} \rightarrow +2CO_{2(aq)} + 2H^+_{(aq)} + 2e^- \quad E^\theta = -0.39 \text{ eV}$$

$$2MnO^-_{4(aq)} + 6H^+_{(aq)} + 5H_2C_2O_{4(aq)} \rightarrow 10CO_{2(aq)} + 2Mn^{2+}_{(aq)} + 4H_2O_{(l)} \quad E^\theta = +1.12 \text{ eV}$$

c) Moles of $MnO_4{}^- = 0.01525 \times 0.02 = 3.05 \times 10^{-4}$ moles. So $2/5 \times (3.05 \times 10^{-4}) = 1.22 \times 10^{-4}$ moles of oxalic acid in 25.00 cm^3. So, concentration $= 1.22 \times 10^{-4}/0.025 = 4.88 \times 10^{-3}$ mol dm^{-3}.

d) As the change from $MnO_4{}^-$ (deep purple) to Mn^{2+} (colourless) is clear from the oxidation state of the manganese, the end point is clearly indicated by one drop of excess $MnO_4{}^-$, which, as it is not reduced, gives the solution a pale pink colour.

e) A chelating agent is a molecule that forms a closed complex in which the ion is removed from solution by forming a multidentate co-ordination complex.

f) The reaction profile may be "S" shaped as the reaction starts slowly, then speeds up a little due to the catalysis effect, and then finally slows down as the reactants run out (see graph on next page).

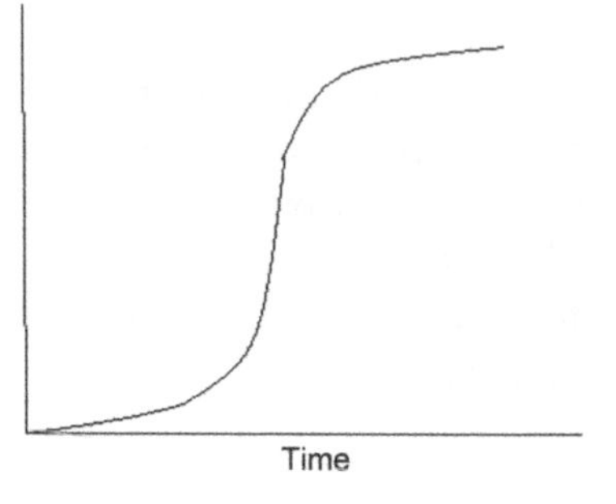

A3.

a) Standard electrode potentials indicate that MnO_2 in acidic conditions are sufficiently powerful oxidizing agents to oxidize both nitrite and ammonium ions to nitrate. For example:

$$MnO_2 + 4H^+ + 2e \rightleftharpoons Mn^{2+} + 2H_2O \quad +1.23$$

$$NO_2^- + H_2O \rightleftharpoons NO_3^- + 2H^+ + 2e \quad -0.94$$

$$MnO_2 + 2H^+ + NO_2^- \rightarrow Mn^{2+} + H_2O + NO_3^- \quad +0.29$$

b) However, both the concentrations of the species in marine environments (very low) and the pH values will mean that the rates of both reactions are also very low. That said, even though the rates are slow, the quantities involved are such that the processes are significant.

A4.

a) $Mn = (Ar)\ 3d^5 4s^2$. $Mn^{2+} = (Ar)\ 3d^5$. The half-filled 3d orbitals are comparatively stable suggesting that the Mn^{2+} ion is stable.

b) $[Mn(H_2O)_6]^{2+} + 2OH^-_{(aq)} \rightarrow [Mn(OH)_2(H_2O)_4](s) + 2H_2O$.

c) This is due to the oxidation of the $[Mn(OH)_2(H_2O)_4](s)$ to MnO_2.

A5.

a) i) $Mn^{2+}_{(aq)} + 2OH^-_{(aq)} \rightarrow Mn(OH)_{2(s)}$. (This is an abbreviated version of the same reaction in A4b.)
ii) $2Mn(OH)_2 + H_2O + \frac{1}{2}O_2 \rightarrow 2Mn(OH)_3$.
iii) $I_2 + 2S_2O_3^{2-} \rightarrow 2I^- + S_4O_6^{2-}$.

b) 8.55 cm^3 of 0.1 M thiosulfate = 8.55×10^{-4} moles. This has reduced 4.275×10^{-4} moles of I_2, which is produced by the reduction of 8.55×10^{-4} moles of Mn^{3+} ions. According to equation (ii) the number of moles of dissolved oxygen is $(0.25)\times8.55\times10^{-4}=2.1375\times10^{-4}$ moles of oxygen in 100 cm^3. Thus, in moles per dm^3 we have 2.14×10^{-3} M equivalent to 6.84 mg dm^{-3}.

c) A volume of 8.55 cm^3 is too low in terms of accuracy. The precision of a burette is ±0.05 cm^3. As two measurements are made before and after the titration a possible error of $0.1/8.55\times100=1.16\%$ is inherent in the procedure. This is reduced if the end point volume is higher.

d) A high dissolved oxygen content can result is increased rates of corrosion of pipes, *etc.* The oxygen content can be removed by heating the water as oxygen's solubility in water decreases with increasing temperature.

A6.

a)
$$\underset{+4}{MnO_{2(s)}} + \underset{-1}{4HCI_{(aq)}} \rightarrow \underset{+2-1}{MnCI_{2(aq)}} + \underset{0}{CI_{2(g)}} + 2H_2O_{(I)}$$

b) 450 kg of $MnO_2 = 4.5\times10^5/87 = 5172.4$ moles. A 1 : 1 ratio means the same number of moles of Cl_2 produced, which occupies $5172.4\times24 = 124\ 137.9$ dm^3 or about 124 m^3.

A7.

a)
$$3\underset{+6}{MnO_4^{2-}} + 4H^+ \rightarrow 2\underset{+7}{MnO_4^-} + \underset{+4}{MnO_2} + 2H_2O$$

b) Anode: $MnO_4{}^{2-} \rightarrow MnO_4{}^- + e^-$.
Cathode: $2H^+ + 2e^- \rightarrow H_2$.

c) In the reaction with acid only two thirds of the manganese is oxidized to manganate VII, leaving MnO_2 as a waste product. The electrolytic method converts 100% of the manganese to manganate VII and also produces hydrogen as a valuable byproduct.

CHAPTER 17

A1.

a) $(0.65\times32/60)+(0.12\times48/100)+(0.11\times128/278)+(0.03\times32/80)+(0.02\times16/40)=0.475$, so in 10 kg of soil, 4.75 kg is oxygen.

b) The only ion likely to give colour would be Fe^{3+}, so a rust/orange colour would be expected.

A2. $2H_2O\rightarrow O_2+4H^++4e^-$ $E^\theta=-1.229$ eV.
$\Delta G^\theta=-4\times(96\,500)\times(-1.229)/1000=+474$ kJ mol^{-1}. The positive ΔG^θ value means that the reaction is not feasible under standard conditions.

A3. A source of chemical energy must allow free energy to be released to enable work to be done. This means that the reactants are thermodynamically unstable with respect to products. However, if the reactants are kinetically unstable, then it is difficult to control the time and place for the reaction to occur. Kinetically stable reactants need sufficient activation energy and/or a catalyst to react.

A4.

a) $(35.73-15.60)=20.13$ g of oxygen.

b) $^{15.6}/_{31}:{}^{20.13}/_{16}$, or 0.503 : 1.26. This is approximately 1 : 2.5, or 2 : 5 P_2O_5.

c) $4Al+3O_2\rightarrow2Al_2O_3$.
12.80 g is $12.8/27=0.474$ moles. From eqn 17.3, $0.474/2=0.237$ moles of Al_2O_3. The mass would be $0.237\times102=24.17$ g.

A5.

a) In the case of both nitrate and sulfate the higher oxidation state of the element forms the acid with the large pK_a.

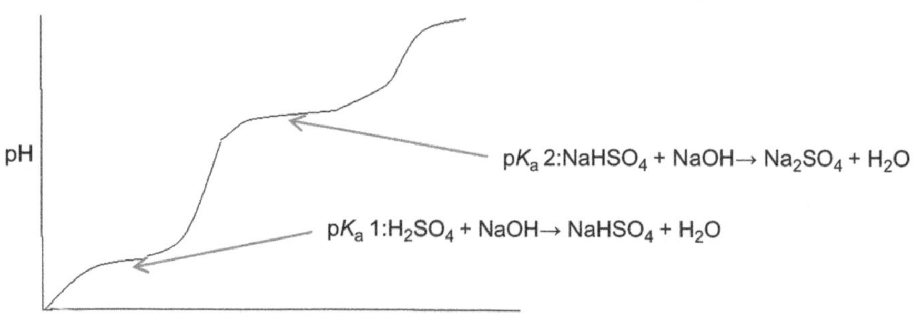

b) A diprotic acid releases two H^+ ions per molecule of acid. This results in two different pK_a values, which can be seen on the graph as the mid-point of the two flatter parts of the curve.
c) Assuming complete ionization at 0.025 M, then a concentration of 0.025 M $H_2SO_4 = 0.05$ M $[H^+]$ ions, which has a pH of $-\log_{10}(0.05) = 1.3$.
d) H_2SO_4: S = +6; H_2SO_3: S = +4.
The extra oxygen results in a greater electron-withdrawing effect from the O-H bonds, causing the O-H bonds to break more easily.

A6.

a) $C_5H_{10}O_5$ $M_R = 150$ g mol^{-1}. %O = 53.3%.
b) 3.
c) $3^2 = 8$ possible optical isomers.
d) D refers to the rotation of plane-polarized light in a clockwise direction by ribose in aqueous solution.

A7.

a) $M_R = 376$ g mol^{-1}. % mass of oxygen = $(6 \times 16)/376 \times 100 =$ 25.5%.
b) Carboxyl/amide functional group and the OH functional group.
c) The π-conjugated electron system in the three-ring system gives rise to absorbances in the visible part of the spectrum. The colour of the solution will be the complementary colour to the wavelengths absorbed.

A8.

a) Vitamin C $M_R = 176$ g mol^{-1}, so $0.09/176 = 5.1 \times 10^{-4}$ moles.
b) Both chiral carbons are identified by the wedge and dashed-line bonds:

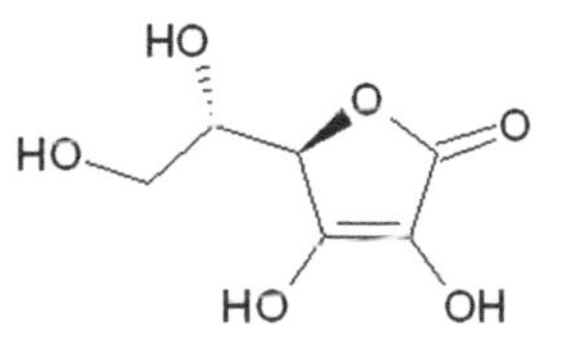

c) Water soluble. The four OH groups would ensure hydrogen bonding with water.

CHAPTER 18

A1. A: Butanone; B: butanal; C: but-2-ene-1-ol (other isomers with same functional groups are possible); D: 1-methoxy propene (other isomers possible); E: cyclobutanol; F: but-1 ene-2-ol.

A2.

a) Compound A, butanone. The C=O group is responsible for the 1720 cm^{-1} absorbance. The methyl group on its own, connected to the C=O group, gives a singlet. On the other side, the ethyl group CH_3CH_2 will give a triplet and a quartet.
b) The isomers are (2*Z*)-but-2-ene-1-ol (left) and (2*E*)-but-2-ene-1-ol (right):

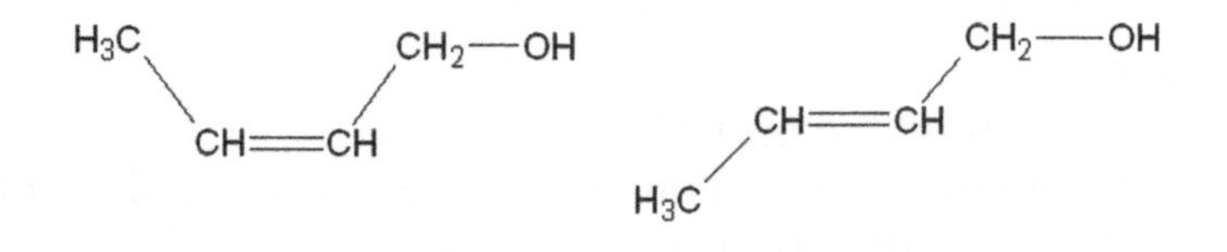

A3. 3.68×10^7!

A4. But-1-ene-2-ol. Chiral carbon indicated by an asterisk (*).

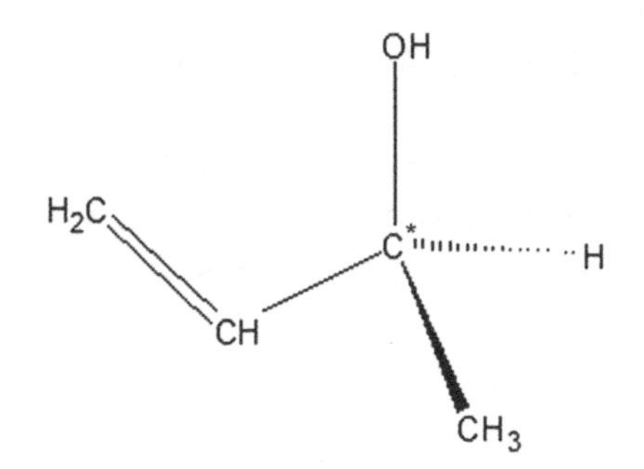

A5.

a) 1 = 2-Amino propanoic acid; 2 = aminoethanoic acid.
b) Only compound 1 shows optical isomerism as the second carbon is attached to four different groups.

A6.

a) pH = 7.0.
b) $C = n/V$; Therefore, $2 = n/(0.05 + 1000)$. $n = 1 \times 10^{-4}$ moles of H_2SO_4, which is 2×10^{-4} moles of H^+ ions.
c) $C = (2 \times 10^{-4})/0.1 = 2 \times 10^{-3} = [H^+]$. Thus, pH = 2.7.

A7.

a) $H_2O + CO_3^{2-} \rightleftharpoons OH^- + HCO_3^- + H^+ \rightleftharpoons H_2O + CO_2$.
These equilibria show how hydrogen carbonate ions can respond to additions of acid or alkali. Hydrogen carbonate ions are the most abundant species of DIC in the oceans (see Fig. 18.2).

b) $CO_{2(g)} + H_2O_{(l)} \rightleftharpoons H_2O_{(l)} + CO_{2(aq)}$. An increase in the partial pressure will push this equilibrium from left to right. Once dissolved in the water, hydrogen carbonate and protons will form (eqn 18.3).

c) pH = 8.2, $[H^+] = 10^{-8.2} = 6.31 \times 10^{-9}$; pH = 8.1, $[H^+] = 10^{-8.1} = 7.94 \times 10^{-9}$. Difference is 7.94–6.31 = 1.63. Expressed as percentage of the original $[H^+] = 1.63/6.31 \times 100 = 25.8\%$.

d) Acidic conditions will significantly decrease the availability of aqueous CO_3^{2-} ions. Organisms that form hard body parts from $CaCO_3$ will find it more difficult.

A8.

a) $K_a = [H^+][HCO_3^-]/[CO_2]$.

b) Rearranging K_a and taking logs: $pH = pK_a + \log(HCO_3^-)/(CO_2)$. So $7.4 = 6.38 + \log(HCO_3^-)/(CO_2)$.
$(HCO_3^-)/(CO_2) = 10^{1.02} = 10.5 : 1$ ratio.

Subject Index